# 1919 G.L.O.
# ADVANCE SHEETS

of Chapters I to VI, inclusive, of a
revison of the
MANUAL OF INSTRUCTIONS
FOR THE
SURVEY OF THE PUBLIC LANDS
OF THE UNITED STATES

## A REPRINT FOR
## CERTIFIED FEDERAL SURVEYORS

By the
General Land Office

Prepared for publication by
Tom Propst, CFedS

DEPARTMENT OF THE INTERIOR
GENERAL LAND OFFICE

# ADVANCE SHEETS

of Chapters I to VI, inclusive, of a
revision of the

# MANUAL OF INSTRUCTIONS

FOR THE

# SURVEY OF THE PUBLIC LANDS OF THE UNITED STATES

Prepared and published under the direction
of the Commissioner of the General Land Office

WASHINGTON
GOVERNMENT PRINTING OFFICE
1919

This is a reprint of an official technical document originally published by Government Printing Office, which is now long out of print.

This document is of particular importance as it gives directions for the layout and retracement of the Public Lands of the United States between the years 1919 and 1930, at which latter date it was superseded by the issuance of an official Manual.

The period between the issuance of the prior Manual in 1902 and these advance sheets of instructions was a time of great change in how the Public Lands were to be surveyed; the end of the contract system of performing surveys and the expansion of legislation regarding the resurvey of the Public Lands.

The tenor of these advance sheets set the tone for all future editions of the Manual.

I hope you find this reprint of some use.

Sincerely,

Tom Propst, P.L.S., CFedS
Irvine, California
Summer, 2011

# DEPARTMENT OF THE INTERIOR
## GENERAL LAND OFFICE

WASHINGTON, *June 16, 1919*.

GENTLEMEN: It has been deemed advisable to publish advance sheets of six chapters of a new edition of the Manual of Surveying Instructions, as follows: (I) Regulations Imposed by Law; (II) Instruments and Methods; (III) System of Rectangular Surveys; (IV) Corner Monuments; (V) Restoration of Lost Corners; and, (VI) Resurveys. These advance sheets will immediately supersede the related provisions of the Manual of 1902, except as may be found impracticable in the case of surveys already in process of execution, or in the instance of returns of surveys now in course of preparation, otherwise the provisions of the Manual of 1902 will remain in full force and effect.

Every member of the surveying service is requested to report to the undersigned any typographical errors which may be detected, to the end that the same may be removed from the completed edition.

Very respectfully,

CLAY TALLMAN,
*Commissioner*.

To the SURVEYING SERVICE OF THE GENERAL LAND OFFICE.

# CONTENTS.

## CHAPTER I.

### Regulations Imposed by Law.



## CHAPTER II.

### Instruments and Methods.

## CHAPTER III.

### SYSTEM OF RECTANGULAR SURVEYS.



## CHAPTER IV.

### CORNER MONUMENTS.

# CHAPTER I.

## REGULATIONS IMPOSED BY LAW.

### THE PUBLIC DOMAIN.

**1.** The survey of the public lands of the United States is inseparably associated with questions relating to the acquisition and disposal of proprietary title to the lands which have been added to the area included in the original thirteen States. The term "public domain" has been applied broadly to the entire aforementioned area in so far as the lands have been subject to survey and disposal by the United States, and of interest herein may be mentioned the twenty-nine States and the District of Alaska surveyed or in progress of survey under the United States rectangular system, as follows:

*Alabama.*—Included in the territory of the original thirteen States, and admitted into the Union December 14, 1819 (3 Stat., 608); surveys practically completed and original records transferred to the Secretary of State at Montgomery.

*Arizona.*—Included in the lands ceded by Mexico, in 1848, and the Gadsden purchase, in 1853, and admitted into the Union February 14, 1912 (36 Stat., 557 and 37 Stat., 1728); surveys in progress; United States Surveyor General at Phoenix.

*Arkansas.*—Acquired under the Louisiana Purchase, in 1803, and admitted into the Union June 15, 1836 (5 Stat., 50); surveys practically completed and original records transferred to the Commissioner of State Lands at Little Rock.

*California.*—Ceded by Mexico, in 1848, and admitted into the Union September 9, 1850 (9 Stat., 452); surveys in progress; United States Surveyor General at San Francisco.

*Colorado.*—Acquired largely under the Louisiana Purchase, in 1803, but including additional land, title to which was quieted through treaty with Spain, in 1819, with other lands annexed with Texas, in 1845, and lands ceded by Mexico, in 1848, and admitted into the Union August 1, 1876 (18 Stat., 474, and 19 Stat., 665); surveys in progress; United States Surveyor General at Denver.

*Florida.*—Ceded by Spain in 1819, and admitted into the Union March 3, 1845 (5 Stat., 742); surveys practically completed and original records transferred to the Commissioner of Agriculture at Tallahassee.

*Idaho.*—Acquired with the Oregon Territory, title to which was established in 1846, and admitted into the Union July 3, 1890 (26 Stat., 215); surveys in progress; United States Surveyor General at Boise.

*Illinois.*—Included in the territory of the original thirteen States and admitted into the Union December 3, 1818 (3 Stat., 536); surveys practically completed and original records transferred to the Auditor of State at Springfield.

*Indiana.*—Included in the territory of the original thirteen States and admitted into the Union December 11, 1816 (3 Stat., 399); surveys practically completed and original records transferred to the Auditor of State at Indianapolis.

*Iowa.*—Acquired under the Louisiana Purchase, in 1803, and admitted into the Union December 28, 1846 (9 Stat., 117); surveys practically completed and original records transferred to the Secretary of State at Des Moines.

*Kansas.*—Acquired under the Louisiana Purchase, in 1803, and with lands annexed with Texas, in 1845, and admitted into the Union January 29, 1861 (12 Stat., 126); surveys practically completed and original records transferred to the Auditor of State and Register of State Lands at Topeka.

*Louisiana.*—Included in the Louisiana Purchase, in 1803, and boundary extended to include additional lands, title to which was quieted through treaty with Spain in 1819, and admitted into the Union April 30, 1812 (2 Stat., 701); surveys practically completed and original records transferred to the Register of State Lands at Baton Rouge.

*Michigan.*—Included in the territory of the original thirteen States and admitted into the Union January 26, 1837 (5 Stat., 144); surveys practically completed and original records transferred to the Commissioner of State Land Office at Lansing.

*Minnesota.*—Included in the territory of the original thirteen States, and with lands acquired under the Louisiana Purchase, in 1803, and admitted into the Union May 11, 1858 (11 Stat., 285); surveys practically completed and original records transferred to the Secretary of State at St. Paul.

*Mississippi.*—Included in the territory of the original thirteen States and admitted into the Union December 10, 1817 (3 Stat., 472); surveys practically completed and original records transferred to the Commissioner of State Lands at Jackson.

*Missouri.*—Acquired under the Louisiana Purchase, in 1803, and admitted into the Union August 10, 1821 (3 Stat., 645, and 3 Stat., Appendix II); surveys practically completed and original records transferred to the Secretary of State at Jefferson City.

*Montana.*—Acquired under the Louisiana Purchase, in 1803, and with the Oregon Territory, title to which was established in 1846, and admitted into the Union November 8, 1889 (25 Stat., 676, and 26 Stat., 1551); surveys in progress; United States Surveyor General at Helena.

*Nebraska.*—Acquired under the Louisiana Purchase, in 1803, and admitted into the Union March 1, 1867 (14 Stat., 391, and 14 Stat., 820); surveys practically completed and original records transferred to the Commissioner of Public Lands and Buildings at Lincoln.

*Nevada.*—Ceded by Mexico in 1848 and admitted into the Union October 13, 1864 (13 Stat., 30, and 13 Stat., 749); surveys in progress; United States Surveyor General at Reno.

*New Mexico.*—Included with lands annexed with Texas, in 1845, with lands ceded by Mexico, in 1848, and the Gadsden Purchase, in 1853, and admitted into the Union January 6, 1912 (36 Stat., 557, and 37 Stat., 1723); surveys in progress; United States Surveyor General at Santa Fe.

*North Dakota.*—Included in the territory of the original thirteen States, and with lands acquired under the Louisiana Purchase, in 1803, and admitted into the Union November 2, 1889 (25 Stat., 676, and 26 Stat., 1548); surveys practically completed and original records transferred to the State Engineer at Bismarck.

*Oklahoma.*—Acquired under the Louisiana Purchase, in 1803, and with lands annexed with Texas, in 1845, and admitted into the Union November 16, 1907 (34 Stat., 267, and 35 Stat., 2160); surveys practically completed and original records filed with the Commissioner of the General Land Office at Washington, D. C.

*Ohio.*—Included in the territory of the original thirteen States, and admitted into the Union April 30, 1802 (2 Stat., 173); surveys practically completed and original records transferred to the Auditor of State at Columbus.

*Oregon.*—Included in the Oregon Territory, title to which was established in 1846, and admitted into the Union February 14, 1859 (11 Stat., 383); surveys in progress; United States Surveyor General at Portland.

*South Dakota.*—Included in the territory of the original thirteen States, and with lands acquired under the Louisiana Purchase, in

1803, and admitted into the Union November 2, 1889 (25 Stat., 676, and 26 Stat., 1549); surveys in progress; United States Surveyor General at Huron.

*Utah.*—Ceded by Mexico in 1848, and admitted into the Union January 4, 1896 (28 Stat., 107, and 29 Stat., 876); surveys in progress; United States Surveyor General at Salt Lake City.

*Washington.*—Included in the Oregon Territory, title to which was established in 1846, and admitted into the Union November 11, 1889 (25 Stat., 676, and 26 Stat., 1552); surveys in progress; United States Surveyor General at Olympia.

*Wisconsin.*—Included in the territory of the original thirteen States, and admitted into the Union May 29, 1848 (9 Stat., 233); surveys practically completed and original records transferred to the Commissioners of Public Lands at Madison.

*Wyoming.*—Included with lands acquired under the Louisiana Purchase, in 1803, with lands annexed with Texas, in 1845, with lands included in the Oregon Territory, title to which was established in 1846, and with lands ceded by Mexico, in 1848, and admitted into the Union July 10, 1890 (26 Stat., 222); surveys in progress; United States Surveyor General at Cheyenne.

*District of Alaska.*—Ceded by Russia in 1867; surveys in progress; United States Surveyor General at Juneau.

**2.** After the admission of the States into the Union the United States continued to hold title to the unappropriated lands and to administer its public-land laws with reference thereto, and it is expressly provided, as one of the conditions set forth in the various enabling acts, that the title to unappropriated lands within the State shall remain in the United States. The lands in the Territories not appropriated by competent authority before they were acquired are in the first instance the exclusive property of the United States, to be disposed of to such persons, at such times, in such modes, and by such titles as the Government may deem most advantageous to the public. Congress alone has the power, derived from Article IV, section 3, of the Constitution, of disposing of the public domain and making all needful rules and regulations in respect thereto.

**3.** Under the laws of the United States the navigable waters have always been and shall forever remain common highways, and below mean high water the same are not subject to survey and disposal. This reservation includes all tidewater streams, and other important permanent bodies of water whose natural and normal condition at the date of the admission of a State into the Union was such as to classify the same as navigable water.

4. The act of Congress approved March 2, 1849 (9 Stat., 352), granted to the State of Louisiana all the swamp and overflowed lands within the limits of the State for the purpose of aiding in the reclamation of said lands, and the act of Congress approved September 28, 1850 (9 Stat., 519), extended the grant to the other public land States then in the Union. The grant was also extended to the States of Minnesota and Oregon by the act of Congress approved March 12, 1860 (12 Stats., 3). The provisions of the aforementioned grants apply to the zone situated below the uplands wherein the lands are of such a character that without the construction of suitable levees and artificial drainage systems the same would be wet and unfit for agricultural purposes. The swamp-land grants apply to all swamp and overflowed lands within the beneficiary States which were unappropriated at the dates of the acts of Congress and whose character *at that time* would bring them within the provisions of said grants. A notable exception to the swamp-land laws is found in the Arkansas Compromise Act approved April 29, 1898 (30 Stat., 367), by virtue of which all right, title, and interest to the remaining unappropriated swamp and overflowed lands within the State of Arkansas reverted to the United States.

5. It comes within the province of the Department of the Interior to consider and determine what are public lands, what lands have been surveyed, what are to be surveyed, what have been disposed of, what remain to be disposed of, and what are reserved, and it is a well-settled principle of law that the United States, through the Department of the Interior, has the right to extend the surveys as may be necessary to include lands omitted from earlier surveys. It is an important duty of the surveyor in the field to discriminate between what are and what are not public lands of the United States and to subdivide the former in accordance with the regulations imposed by law.

## LAWS RELATING TO SURVEYS.

6. The rectangular surveying system is based upon existing law and was devised with the object of marking upon the ground and fixing for all time legal subdivisions for purposes of description and disposal of the public domain under the general land laws of the United States.

7. The rectangular system of survey of the public lands was inaugurated by a committee appointed by the Continental Congress.

On the 7th of May, 1784, this committee reported "An ordinance for ascertaining the mode of locating and disposing of lands in the western territory, and for other purposes therein mentioned." The ordinance as finally passed on the 20th of May, 1785, provided for townships 6 miles square, containing 36 sections of 1 mile square. The first public surveys were made under this ordinance. The townships, 6 miles square, were laid out in ranges extending northward from the Ohio River, the townships being numbered from south to north, and the ranges from east to west. The region embraced by the surveys under this law forms a part of the State of Ohio. In these initial surveys only the exterior lines of the townships were surveyed, but the plats were marked by subdivisions into sections of 1 mile square, and mile corners were established on

| | | | | | |
|---|---|---|---|---|---|
| 6 | 5 | 4 | 3 | 2 | 1 |
| 7 | 8 | 9 | 10 | 11 | 12 |
| 18 | 17 | 16 | 15 | 14 | 13 |
| 19 | 20 | 21 | 22 | 23 | 24 |
| 30 | 29 | 28 | 27 | 26 | 25 |
| 31 | 32 | 33 | 34 | 35 | 36 |

the township lines. The sections were numbered from 1 to 36, and the surveys were made under the direction of the Geographer of the United States.

The act of Congress approved May 18, 1796, provided for the appointment of a surveyor general and directed the survey of the lands northwest of the Ohio River and above the mouth of the Kentucky River, "in which the titles of the Indian tribes have been extinguished." Under this law it was provided that "the sections shall be numbered, respectively, beginning with the number one in the northeast section and proceeding west and east alternately through the township, with progressive numbers till the thirty-sixth be completed." This method of numbering sections, as shown by the accompanying diagram, is still in use.

The act of Congress approved May 10, 1800, required the "townships west of the Muskingum, which * * * are directed to be sold in quarter townships, to be subdivided into half sections of three hundred and twenty acres each, as nearly as may be, by running parallel lines through the same from east to west and from south to north at the distance of one mile from each other, and marking corners at the distance of each half mile on the lines running from east to west and at the distance of each mile on those running from south to north. * * * And the interior lines of townships intersected by the Muskingum, and of all the townships lying east of that river, which have not been heretofore actually subdivided into sections shall also be run and marked. * * * And in all cases where the exterior lines of the townships thus to be subdivided into sections or half sections shall exceed, or shall not extend, six miles, the excess or deficiency shall be specially noted and added to or deducted from the western and northern ranges of sections or half sections in such townships, according as the error may be in running the lines from east to west or from south to north."

The act of Congress approved February 11, 1805, directs the subdivision of the public lands into quarter sections and provides that all the corners marked in the public surveys shall be established as the proper corners of sections, or subdivisions of sections, which they were intended to designate, and that corners of half and quarter sections not marked shall be placed as nearly as possible "equidistant from those two corners which stand on the same line." This act further provides that "The boundary lines actually run and marked * * * shall be established as the proper boundary lines of the sections or subdivisions for which they were intended; and the length of such lines as returned by * * * the surveyors * * * shall be held and considered as the true length thereof, and the boundary lines which shall not have been actually run and marked as aforesaid shall be ascertained by running straight lines from the established corners to the opposite corresponding corners; but in those portions of the fractional townships where no such opposite or corresponding corners have been or can be fixed, the said boundary lines shall be ascertained by running from the established corners due north and south or east and west lines, as the case may be, to the * * * external boundary of such fractional township."

The act of Congress approved April 25, 1812, provided "That there shall be established in the Department of the Treasury an office to be denominated the General Land Office, the chief officer of which shall be called the Commissioner of the General Land Office, whose duty it shall be, under the direction of the head of the department, to superintend, execute, and perform all such acts and things touching or respecting the public lands of the United States, and other lands patented or granted by the United States, as have heretofore been directed by law to be done or performed in the office of the Secretary of State, of the Secretary and Register of the Treasury, and of the Secretary of War, or which shall hereafter by law be assigned to the said office."

The act of Congress approved April 24, 1820, provides for the sale of public lands in half-quarter sections, and requires that "in every case of the division of a quarter section the line for the division thereof shall run north and south * * * and fractional sections, containing one hundred and sixty acres and upward, shall, in like manner, as nearly as practicable, be subdivided into half-quarter sections, under such rules and regulations as may be prescribed by the Secretary of the Treasury; but fractional sections containing less than one hundred and sixty acres shall not be divided."

The act of Congress approved May 29, 1830 (secs. 2412, 2413, R. S.), provides for the fine and imprisonment of any person obstructing the survey of the public lands, and for the protection of surveyors, in the discharge of their official duties, by the United States marshal, with sufficient force, whenever necessary.

The act of Congress approved April 5, 1832, directed the subdivision of the public lands into quarter quarters; that in every case of the division of a half-quarter section the dividing line should run east and west; and that fractional sections should be subdivided under rules and regulations prescribed by the Secretary of the Treasury. Under the latter provision the Secretary directed that fractional sections containing less than 160 acres, or the residuary portion of a fractional section, after the subdivision into as many quarter-quarter sections as it is susceptible of, may be subdivided into lots, each containing the quantity of a quarter-quarter section as nearly as practicable, by so laying down the line of subdivision that they shall be 20 chains wide, which distances are to be marked on the plat of subdivision, as are also the areas of the quarter quarters and residuary fractions.

The last two acts above mentioned provided that the corners and contents of half-quarter and quarter-quarter sections should be ascertained, as nearly as possible, in the manner and on the principles directed and prescribed in the act of Congress approved February 11, 1805.

The act of Congress approved July 4, 1836, provided for the reorganization of the General Land Office, and that the executive duties of said office "shall be subject to the supervision and control of the Commissioner of the General Land Office under the direction of the President of the United States." The repealing clause is, "That such provisions of the act of the twenty-fifth of April, in the year one thousand eight hundred and twelve, entitled 'An act for the establishment of a General Land Office in the Department of the Treasury,' and of all acts amendatory thereof, as are inconsistent with the provisions of this act, be, and the same are hereby, repealed."

From the wording of this act it would appear that the control of the General Land Office was removed from the Treasury Department, and that the commissioner reported directly to the President; but, as a matter of fact, the Secretary of the Treasury still had supervisory control, for the act of Congress approved March 3, 1849, by which the Department of the Interior was established, provided, "That the Secretary of the Interior shall perform all the duties in relation to the General Land Office, of supervision and appeal, now discharged by the Secretary of the Treasury * * *." By this act the General Land Office was transferred to the Department of the Interior, where it still remains.

8. The following comprises so much of the general laws relating to the survey of the public domain as it is deemed necessary to incorporate in this volume, reference being made by chapter and section to the codification of the Public Land Laws, prepared pursuant to acts of Congress approved March 3, 1879, and June 16, 1880, and by section number to the Revised Statutes of the United States.

Duties of Commissioner.

SEC. 32. The Commissioner of the General Land Office shall perform, under the direction of the Secretary of the Interior, all executive duties appertaining to the surveying and sale of the public lands of the United States, or in anywise respecting such public lands; and, also, such as relate to private claims of lands, and the issuing of patents for all grants of land under the authority of the Government. (R. S., 453.)

Power of Commissioner to make regulations.

SEC. 61. The Commissioner, under the direction of the Secretary of the Interior, is authorized to enforce and carry into execution every part of the public land laws not otherwise specially provided for. (R. S., 2478.)

Surveyor general, how and where appointed.

SEC. 77. There shall be appointed by the President, by and with the advice and consent of the Senate, a surveyor general for the States and Territories herein named, embracing, respectively, one surveying district, namely: Louisiana, Florida, Minnesota, Kansas, California, Nevada, Oregon, Nebraska and Iowa, Dakota, Colorado, New Mexico, Idaho, Washington, Montana, Utah, Wyoming, Arizona. (R. S., 2207.)

Residence of surveyor general.

SEC. 83. Every surveyor general, while in the discharge of the duties of his office, shall reside in the district for which he is appointed. (R. S. 2214.)

Bond of surveyor general.

SEC. 84. Every surveyor general shall, before entering on the duties of his office, execute and deliver to the Secretary of the Interior a bond, with good and sufficient security, for the penal sum of thirty thousand dollars, conditioned for the faithful disbursement, according to law, of all public money placed in his hands, and for the faithful performance of the duties of his office; and the President has discretionary authority to require a new bond and additional security, under the direction of the Secretary of the Interior, for the lawful disbursement of public moneys. (R. S., 2215, 2216.)

Duration of office.

SEC. 85. The commission of each surveyor general shall cease and expire in four years from the date thereof, unless sooner vacated by death, resignation, or removal from office. (R. S., 2217.)

Continuance of duties and bond after expiration of commission.

SEC. 86. Every surveyor general, except where the President sees cause otherwise to determine, is authorized to continue in the uninterrupted discharge of his regular official duties after the day of expiration of his commission and until a new commission is issued to him for the same office, or until the day when a successor enters upon the duties of such office; and the existing official bond of any officer so acting shall be deemed good and sufficient and in force until the date of the approval of the new bond to be given by him, if recommissioned, or otherwise, for the additional time he may so continue

officially to act, pursuant to the authority of this section. (R. S., 2222.)

SEC. 87. Whenever the surveys and records of any surveying district are completed the surveyor general thereof shall be required to deliver over to the secretary of state of the respective States, including such surveys, or to such other officer as may be authorized to receive them, all the field notes, maps, records, and other papers appertaining to land titles within the same; and the office of surveyor general in every such district shall thereafter cease and be discontinued. (R. S., 2218.)

Transfer of papers and discontinuance of office in case of completed surveys.

SEC. 88. In all cases of discontinuance, as provided in the preceding section, the authority, powers, and duties of the surveyor general in relation to the survey, resurvey, or subdivision of the lands therein, and all matters and things connected therewith, shall be vested in and devolved upon the Commissioner of the General Land Office. (R. S., 2219.)

Devolution of powers upon Commissioner in case of discontinuance.

SEC. 89. Under the authority and direction of the Commissioner of the General Land Office any deputy surveyor or other agent of the United States shall have free access to any such field notes, maps, records, and other papers for the purpose of taking extracts therefrom or making copies thereof without charge of any kind; but no transfer of such public records shall be made to the authorities of any State until such State has provided by law for the reception and safekeeping of such public records, and for the allowance of free access thereto by the authorities of the United States. (R. S., 2220, 2221.)

Free access to public records delivered to States, and condition of such delivery.

SEC. 99. First. The public lands shall be divided by north and south lines run according to the true meridian, and by others crossing them at right angles, so as to form townships of six miles square, unless where the line of an Indian reservation, or of tracts of land heretofore surveyed or patented, or the course of navigable rivers, may render this impracticable; and in that case this rule must be departed from no further than such particular circumstances require.

Rules of survey.

Second. The corners of the townships must be marked with progressive numbers from the beginning; each distance of a mile between such corners must be also distinctly marked with marks different from those of the corners.

Third. The township shall be subdivided into sections, containing as nearly as may be, six hundred and forty acres each, by running through the same, each way, parallel lines at the end of every two miles[1]; and by making a corner on each of such lines at the end of every mile. The sections shall be numbered, respectively, beginning with the number one in the northeast section, and proceeding west and east alternately through the township with progressive numbers till the thirty-six be completed.

Fourth. The deputy surveyors, respectively, shall cause to be marked on a tree near each corner established in the manner described, and within the section, the number of such section, and over it the number of the township within which such section may be; and the deputy surveyors shall carefully note, in their respective field books, the names of the corner trees marked and the numbers so made.

Fifth. Where the exterior lines of the townships which may be subdivided into sections or half sections exceed, or do not extend six miles, the excess or deficiency shall be specially noted, and added to or deducted from the western and northern ranges of sections or half sections in such townships, according as the error may be in running the lines from east to west, or from south to north; the sections and half sections bounded on the northern and western lines of such townships shall be sold as containing only the quantity expressed in the returns and plats, respectively, and all others as containing the complete legal quantity.

Sixth. All lines shall be plainly marked upon trees, and measured with chains, containing two perches of sixteen and one-half feet each, subdivided into twenty-five equal links; and the chain shall be adjusted to a standard to be kept for that purpose.[2]

Seventh. Every surveyor shall note in his field book the true situations of all mines, salt licks, salt springs, and mill seats which come to his knowledge; all water courses over which the line he runs may pass; and also the quality of the lands.

Eighth. These field books shall be returned to the surveyor general, who shall cause therefrom a description of the whole lands sur-

---

1 Authority for the establishment of section lines at intervals of 1 mile is found in the act of Congress approved May 10, 1800, previously quoted.

2 The superior results obtained by the use of modern steel ribbon tapes, in contrast with the obsolete link chain, have led to the abandonment of the latter, except that the "chain unit," which is peculiarly adapted to land surveying, has always been employed.

veyed to be made out and transmitted to the officers who may superintend the sales. He shall also cause a fair plat to be made of the townships and fractional parts of townships contained in the lands, describing the subdivisions thereof, and the marks of the corners. This plat shall be recorded in books to be kept for that purpose; and a copy thereof shall be kept open at the surveyor general's office for public information, and other copies shall be sent to the places of the sale and to the General Land Office. (Acts of May 18, 1796, and May 10, 1800, and R. S., 2395.)

Boundaries and contents of public lands, how ascertained.

SEC. 100. The boundaries and contents of the several sections, half sections, and quarter sections of the public lands shall be ascertained in conformity with the following principles:

First. All the corners marked in the surveys returned by the surveyor general shall be established as the proper corners of sections, or subdivisions of sections, which they were intended to designate, and the corners of half and quarter sections, not marked on the surveys, shall be placed as nearly as possible equidistant from two corners which stand on the same line.

Second. The boundary lines, actually run and marked in the surveys returned by the surveyor general, shall be established as the proper boundary lines of the sections or subdivisions for which they were intended, and the length of such lines as returned shall be held and considered as the true length thereof. And the boundary lines which have not been actually run and marked shall be ascertained by running straight lines from the established corners to the opposite corresponding corners; but in those portions of the fractional townships, where no such opposite corresponding corners have been or can be fixed, the boundary lines shall be ascertained by running from the established corners due north and south or east and west lines, as the case may be, to the water course, Indian boundary line, or other external boundary of such fractional township.

Third. Each section or subdivision of section, the contents whereof have been returned, by the surveyor general, shall be held and considered as containing the exact quantity expressed in such return; and the half sections and quarter sections, the contents whereof shall not have been thus returned, shall be held and considered as containing the one-half or the one-fourth part, respectively, of the returned contents of the section of which they may make part. (Act of Feb. 11, 1805, and R. S., 2396.)

Lines of division of half-quarter sections, how run.

SEC. 101. In every case of the division of a quarter section the line for the division thereof shall run north and south, and the corners and contents of half-quarter sections which may thereafter be sold shall be ascertained in the manner and on the principles directed and prescribed by the section preceding, and fractional sections containing one hundred and sixty acres or upwards shall in like manner, as nearly as practicable, be subdivided into half-quarter sections, under such rules and regulations as may be prescribed by the Secretary of the Interior, and in every case of a division of a half-quarter section, the line for the division thereof shall run east and west, and the corners and contents of quarter-quarter sections, which may thereafter be sold, shall be ascertained, as nearly as may be, in the manner and on the principles directed and prescribed by the section preceding; and fractional sections containing fewer or more than one hundred and sixty acres shall in like manner, as nearly as may be practicable, be subdivided into quarter-quarter sections, under such rules and regulations as may be prescribed by the Secretary of the Interior. (R. S., 2397.)

Extension of public surveys over mineral lands.

SEC. 106. The public surveys shall extend over all mineral lands, and all subdividing of surveyed lands into lots less than one hundred and sixty acres may be done by county and local surveyors at the expense of claimants; but nothing in this section contained shall require the survey of waste or useless lands. (R. S., 2406.)

Surveyors general to survey private land claims when confirmed, etc.

SEC. 118. Each surveyor general, when thereunto duly authorized by law, shall cause all confirmed private land claims within his district to be accurately surveyed, and shall transmit plats and field notes thereof to the Commissioner of the General Land Office for his approval. When publication of such surveys is authorized by law, the proof thereof, together with any objections properly filed, and all evidence submitted either in support of or in opposition to the approval of any such survey, shall also be transmitted to said Commissioner. (R. S., 2447.)

Penalty for interrupting surveys.

SEC. 120. Every person who in any manner, by threat or force, interrupts, hinders, or prevents the surveying of the public lands, or of any private land claim which has been or may be confirmed by the United States, by the persons authorized to survey the same, in conformity with the instructions of the Commissioner of the General Land Office, shall be fined not less than fifty dollars, nor more than three thousand dollars, and be imprisoned not less than one nor more than three years. (R. S., 2412.)

Protection of surveyor by marshal of district.

SEC. 121. Whenever the President is satisfied that forcible opposition has been offered, or is likely to be offered, to any surveyor or deputy surveyor in the discharge of his duties in surveying the public lands, it may be lawful for the President to order the marshal of the State or district, by himself or deputy, to attend such surveyor or deputy surveyor with sufficient force to protect such officer in the execution of his duty, and to remove force should any be offered. (R. S., 2413.)

9. More recent legislation has brought about (*a*) provision for the appointment of a United States Surveyor General for the District of Alaska; (*b*) authority for the purchase of durable monuments, to be employed in place of native material to mark public land corners; (*c*) penalty for the destruction of monuments of the public land surveys; (*d*) authority for necessary resurveys; and (*e*) change of survey system from contract to direct with authority for the employment of a permanent corps of United States surveyors; all as indicated by the following quotations from the United States Statutes:

Surveyor general for the District of Alaska.

The act of Congress approved May 17, 1884, providing a civil government for Alaska, provides "That the said District of Alaska is hereby created a land district, * * * and the marshal provided for by this act shall be ex officio surveyor general of said district." (23 Stat., 24, sec. 8.) The act of Congress approved July 24, 1897, amends the act approved May 17, 1884, and provides "That there shall be appointed by the President, by and with the advice and consent of the Senate, a surveyor general for the District of Alaska, embracing one surveying district." (30 Stat., 215, sec. 2.)

Purchase of metal monuments.

The act of Congress approved May 27, 1908, provided "for the purchase of metal monuments to be used for public land survey corners wherever practicable." (35 Stat., 347.) This authority was amplified by the act of Congress approved June 25, 1910, making appropriation for sundry civil expenses for the fiscal year ended June 30, 1911, and has been continued from year to year to the present time. The act approved July 1, 1918, provided, under "Surveying the Public Lands," as follows:

"That the sum of not exceeding 10 per centum of the amount hereby appropriated may be expended by the Commissioner of the General Land Office, with the approval of the Secretary of the Interior, for the purchase of metal or other equally durable monuments to be used for public land survey corners wherever practicable: * * *." (40 Stat., 668.)

Penalty for the destruction of survey monuments.

The act of Congress approved March 4, 1909, entitled "An act to codify, revise, and amend the penal laws of the United States," provides punishment for offenses against the operation of the surveying service of the Government, as follows:

"Whoever shall willfully destroy, deface, change, or remove to another place any section corner, quarter-section corner, or meander post, on any Government line of survey, or shall willfully cut down any witness tree or any tree blazed to mark the line of a Government survey, or shall willfully deface, change, or remove any monument or bench mark of any Government survey, shall be fined not more than $250, or imprisoned not more than six months, or both." (35 Stat., 1088, sec. 57.)

Resurvey of public lands.

The act of Congress approved March 3, 1909, entitled "An act authorizing the necessary resurvey of public lands," as amended by joint resolution approved June 25, 1910, provides as follows:

"That the Secretary of the Interior may, in his discretion, cause to be made, as he may deem wise under the rectangular system now provided by law, such resurveys or retracements of the surveys of public lands as, after full investigation, he may deem essential to properly mark the boundaries of the public lands remaining undisposed of: *Provided*, That no such resurvey or retracement shall be so executed as to impair the bona fide rights or claims of any claimant, entryman, or owner of lands affected by such resurvey or retracement: *Provided further*, That not to exceed 20 per centum of the total annual appropriation for surveys and resurveys of the public lands shall be used for the resurveys and retracements authorized hereby." (35 Stat., 845, and 36 Stat., 884.)

The act of Congress approved September 21, 1918, entitled "An act authorizing the resurvey or retracement of lands heretofore returned as surveyed public lands of the United States under certain conditions", provides authority for the resurvey by the Government of townships heretofore held to be ineligible for resurvey under existing regulations of the Department of the Interior by reason of disposals in excess of fifty per centum of the total area thereof. The act provides:

"That upon the application of the owners of three-fourths of the privately owned lands in any township covered by public-land surveys, more than fifty per centum of the area of which townships is privately owned, accompanied by a deposit with the United States surveyor general for the proper State, or if there be no surveyor general of such State, then with the Commissioner of the General Land Office, of the proportionate estimated cost, inclusive of the necessary (office) work, of the resurvey or retracement of all the privately owned lands in said township, the Commissioner of the General Land Office, subject to the supervisory authority of the Secretary of the Interior, shall be authorized in his discretion to cause to be made a resurvey or retracement of the lines of said township and to set permanent corners and monuments in accordance with the laws and regulations governing surveys and resurveys of public lands; that the sum so deposited shall be held by the surveyor general or commissioner when ex officio surveyor general and may be expended in payment of the cost of such survey, including field and office work, and any excess over the cost of such survey and the expenses incident thereto shall be repaid pro rata to the persons making said deposits or their legal representatives; that the proportionate cost of the field and office work for the resurvey or retracement of any public lands in such township shall be paid from the current appropriation for the survey and resurvey of public lands, in addition to the portion of such appropriation otherwise allowed by law for resurveys and retracements; that similar resurveys and retracements may be made on the application, accompanied by the requisite deposit, of any court of competent jurisdiction, the returns of such resurvey or retracement to be submitted to the court; that the Secretary of the Interior is authorized to make all necessary rules and regulations to carry this act into full force and effect." (40 Stat., 965.)

**Selection of surveyors.**

The act of Congress approved June 25, 1910 (36 Stat., 703, 740), making appropriation for sundry civil expenses for the fiscal year ended June 30, 1911, provided, under "Surveying the Public Lands": "The surveys and resurveys to be made by such competent surveyors as the Secretary of the Interior may select, * * *." This provision of law brought to a close the contract system which had theretofore been adhered to since the beginning of the public land surveys, and the authority for the employment of a permanent corps of United States surveyors has been continued from year to year to the present time. The following comprises that part of the act of Congress approved July 1, 1918, under "Surveying the Public Lands," relating directly to the administrative control of the surveying service:

"For surveys and resurveys of public lands, under the supervision of the Commissioner of the General Land Office and direction of the Secretary of the Interior, * * *. The surveys and resurveys provided for in this appropriation to be made by such competent surveyors as the Secretary of the Interior may select, * * *." (40 Stat., 668.)

## GENERAL RULES.

**10.** From the foregoing synopsis of congressional legislation it is evident—

First. That the boundaries of the public lands established and returned by the duly appointed surveyors, when approved by the surveyors general and accepted by the Commissioner of the General Land Office, are unchangeable.

Second. That the original township, section, and quarter-section corners established by the surveyors must stand as the true corners which they were intended to represent, whether in the place shown by the field notes or not.

Third. That quarter-quarter-section corners not established in the process of the original survey shall be placed on the line connecting the section and quarter-section corners, and midway between them, except on the last half mile of section lines closing on the north and west boundaries of the township, or on other lines between fractional or irregular sections.

Fourth. That the center lines of a regular section are to be straight, running from the quarter-section corner on one boundary of the section to the corresponding corner on the opposite section line.

Fifth. That in a fractional section where no opposite corresponding quarter-section corner has been or can be established, the center line of such section must be run from the proper quarter-section corner as nearly in a cardinal direction to the meander line, reservation or other boundary of such fractional section, as due parallelism with section lines will permit.

ꞏliterated corners of the approved surveys eir original locations whenever it is possible to decisions by surveyors which may result in ꞏaries of patented lands and disturb questions of connection therewith are subject to review by the

THE MANUAL.

ꞏarious regions of the United States have been surveyed ꞏr different sets of instructions issued at periods ranging from 85 to the present time. The earliest rules were given to surveyors in manuscript or in printed circulars. Regulations more in detail, improving the system for greater accuracy, permanency and uniformity, were issued in book form in editions of 1855, 1881, 1890, 1894 and 1902.

The Manual of Surveying Instructions has been again revised with a view to harmonizing the printed instructions furnished to the surveyors with recent legislation and current surveying practice. The use of iron-post corner monuments adds much to the permanency of the evidence of the surveys, but this calls for little change in rules except to outline the standard practice. A growing necessity for resurveys to identify and restore original surveys actually made, but poorly monumented, or to supersede grossly erroneous or fraudulent original surveys—"to properly mark the boundaries of the public land remaining undisposed of"—has demanded a full discussion of the subject in this revision of the Manual. The change from the contract system to the present system under which the public-land surveys are executed by a permanent corps of surveyors employed by the General Land Office has involved changes in the administrative control without departing from previous technical procedure, and hereafter throughout the Manual all reference to administrative questions will be found to be stated in general terms in order to avoid confusing that matter with the purely technical subjects. Modern surveying practice has been introduced into the public-land surveys as far as legally consistent and efficient, which has prompted a rather full instructive treatment of the subjects of measurements with long steel tapes, stadia method and triangulations, and field observations for the determination of time, latitude and azimuth, to afford versatility on the part of the surveyor in adopting methods best suited to the ever-changing conditions under which his work must be accomplished.

The instructions contained in this Manual are to be observed by every surveyor engaged in the execution of the public-land surveys. All other surveyors, including those who have at times been employed in the surveying service of the General Land Office, should bear in mind that in their private capacities they are acting under somewhat different rules of law from those governing original surveys, and surveyors should discriminate between the provisions of the statute which control original surveys and those which apply to the retracement of lines that have been officially established and approved.

## THE STANDARD FIELD TABLES.

**12.** There has been published by the General Land Office, in the shape of a pocket field book, a compendium of tables and formulas entitled "Standard Field Tables." The volume embraces the data peculiarly useful to surveyors engaged in subdividing the public lands. The Standard Field Tables are issued as a supplement to the Manual, and as such the former are a part of the latter, with contents as follows:

1. Units of linear measure, units of area, expansion of steel tapes, and conversion tables; chains to feet and feet to chains.
2. Reduction in latitude to south boundary of township, and corrections for convergency within a township.
3. Traverse table, for the correction of random lines.
4. Traverse tables.
5. Correction of error in stadia wire interval.
6. Stadia coefficients, vertical rod.
7. Natural sines and cosines.
8. Natural tangents and cotangents.
9. Logarithmic sines, cosines, tangents and cotangents.
10. Logarithms of numbers.
11. Convergency of meridians, and differences of latitude and longitude.
12. Azimuths of the tangent to the parallel.
13. Offsets from the tangent to the parallel.
14. Azimuths of the secant.
15. Offsets from the secant to the parallel.
16. Lengths of arcs of the earth's surface.
17. Apparent time of sunrise and sunset.
18. Conversion tables, degrees to time, and time to degrees.
19. Sidereal conversions, and reductions to the local mean time of upper culmination of Polaris.

20. Mean refractions in zenith distance.
21. Coefficients to apply to mean refractions for variations in barometer and temperature.
22. Coefficients for computing errors in azimuth due to small errors in declination or latitude.
23. Mean refractions in polar distance.
24. Trigonometric formulas for the solution of plane triangles.
25. Trigonometric formulas for the solution of stadia measurements, observations for time, latitude and azimuth, and problems in convergency.

### EPHEMERIS OF THE SUN AND POLARIS, AND TABLES OF AZIMUTHS OF POLARIS.

**13.** The above title has been given to a second supplement to the Manual which is published each year, a convenience which serves to supply the surveyors with all necessary data relating to the daily positions of the sun and Polaris without requiring frequent revision of the text of the Manual or the Standard Field Tables. As a supplement to the Manual the data contained in the Ephemeris will be adopted in preference to that contained in other publications over which the General Land Office has no control either as to accuracy or fitness for use in the public-land surveys.

# CHAPTER II.

## INSTRUMENTS AND METHODS.

### MEASUREMENTS.

**14.** The law prescribes the chain as the unit of linear measure for the survey of the public lands, and all returns of measurements are to be made in true horizontal distances, in miles, chains and links. The chain unit is known as the invention of Edmund Gunter, an English astronomer of the seventeenth century, and is especially convenient in computing areas in the unit of acres, one acre being equal to 10 square chains.

*Units of linear measure.*

1 chain=100 links.
=66 feet.
1 mile=80 chains.
=5,280 feet.

*Units of area.*

1 acre=10 square chains.
=43,560 square feet.
1 square mile=640 acres.

**15.** Each surveyor will be provided with a standard and an assortment of 1, 2, 5 or 8-chain steel tapes. The standard tape will be employed for comparison with the field tapes, in order that errors in the latter may be noted and corrected. Before chainmen are intrusted with their actual duties they should be instructed by the chief of party, and required to measure over one or more trial lines of level and mountainous surface, to secure accuracy and uniformity of results.

**16.** It is essential to the record of a survey to state briefly at the beginning of the field notes, with every set of returns, the general manner of making measurements in the survey, and as topographical difficulties are encountered making it necessary to depart from the stated general method, it is desirable to record the plan of special measurement adopted. The field notes thus exhibit the manner of making all measurements, and the record should be such that another surveyor retracing any line can substantially duplicate the exact procedure adopted in the original survey.

The following paragraphs are illustrative of the record to be made in the field notes:

"Unless otherwise specified all measurements are made with a Chicago 1-chain steel tape compared with a Chesterman standard steel tape and found correct."

"Unless otherwise specified all measurements are made with a Lallie 2-chain steel tape found correct by comparison with a Lufkin standard steel tape."

"Unless otherwise specified all measurements are made with a Lufkin steel tape 8 chains in length compared with a Chesterman standard steel tape and found correct. The measurements are made on the slope, the vertical angle determined, and the slope measurements properly reduced to true horizontal distances."

### THE LONG STEEL TAPE.

**17.** The most approved method of measurement involves the use of steel ribbon tapes from 2 to 8 chains in length; in its use in the public-land surveys the tape is properly alined and stretched, and the measurements are made on the slope at any convenient distance up to the length of the tape as limited by the topography. The vertical angles of the lesser slopes are determined by the use of clinometers in the hands of the chainmen, while the vertical angles of the particularly sharp slopes are determined with the transit operated by the surveyor. The slope distances are then reduced to true horizontal distances and the entire operation suitably recorded. It is not considered necessary to exhibit in the official field notes any but the true horizontal distances, omitting details, except where precise measurements are made of various bases for special use.

**18.** The following is an example of both field and final record for the use of the long steel tape and clinometer, and reductions by the use of the traverse tables (see Table 4, Standard Field Tables):

| Field record. | | | | | Final field notes. | |
|---|---|---|---|---|---|---|
| Mean vertical angle. | Distance on slope. | True horizontal distance. | Intermediate measurement. | Difference in elevation. | | |
| | *Chains.* | *Chains.* | *Chains.* | *Chains.* | *Chains.* | North, bet. secs. 19 and 24. Desc. 155 ft. over NW. slope, through scattering timber and dense undergrowth. |
| $-12\frac{1}{4}°$ | 4.50 | 4.398 | | −0.95 | | |
| $-17\frac{1}{2}°$ | 2.20 | 2.098 | | − .66 | | |
| | 6.70 | 6.496 | 3.80 | − .75 | 10.30 | Dry gulch, course W.; asc. 295 ft. over SW. slope. |
| | | | | + .75 | | |
| $+8\frac{1}{4}°$ | 8.00 | 7.917 | | +1.15 | | |
| | 14.70 | 14.413 | | | | |
| $+19\frac{3}{4}°$ | 6.20 | 5.835 | | +2.10 | | |
| | 20.90 | 20.248 | | | | |
| $+7\frac{3}{4}°$ | 3.30 | 3.270 | | + .44 | | |
| | 24.20 | 23.518 | 0.00 | | 23.50 | Spur, slopes W.; desc. 185 ft. to 1/4 sec. cor., over NW. slope. |
| | | | 1.20 | | 24.70 | Wagon road, bears E. and W. |
| | | | 1.90 | | 25.40 | Leave undergrowth. |
| $-6\frac{1}{2}°$ | 8.00 | 7.949 | | −0.91 | | |
| | 32.20 | 31.467 | 1.15 | | 32.60 | Enter heavy timber, bears NW. and SE. |
| $-10\frac{1}{4}°$ | 3.70 | 3.641 | | − .66 | | |
| | 35.90 | 35.108 | | | | |
| −14 ° | 5.00 | 4.851 | | −1.21 | | |
| | 40.90 | 39.959 | | | | |
| 0 ° | .04 | .04 | | | | |
| | 40.94 | 40.00 | | | | |
| | 40.00 | 40.00 | | | 40.00 | Set an iron post, etc. |

**19.** A simplification of the reduction of measurements on the slope is obtained by the use of two diagrams constructed on cross-section paper, as follows: The first with the vertical lines representing intervals of 20 links measurement on the slope to 2, 5 or 8 chains to suit the length of tape used; the horizontal lines representing the correction in links to be made from the measurement on the slope to obtain the true horizontal distance; slanting lines are drawn to represent various degrees of slope scaled to the proper

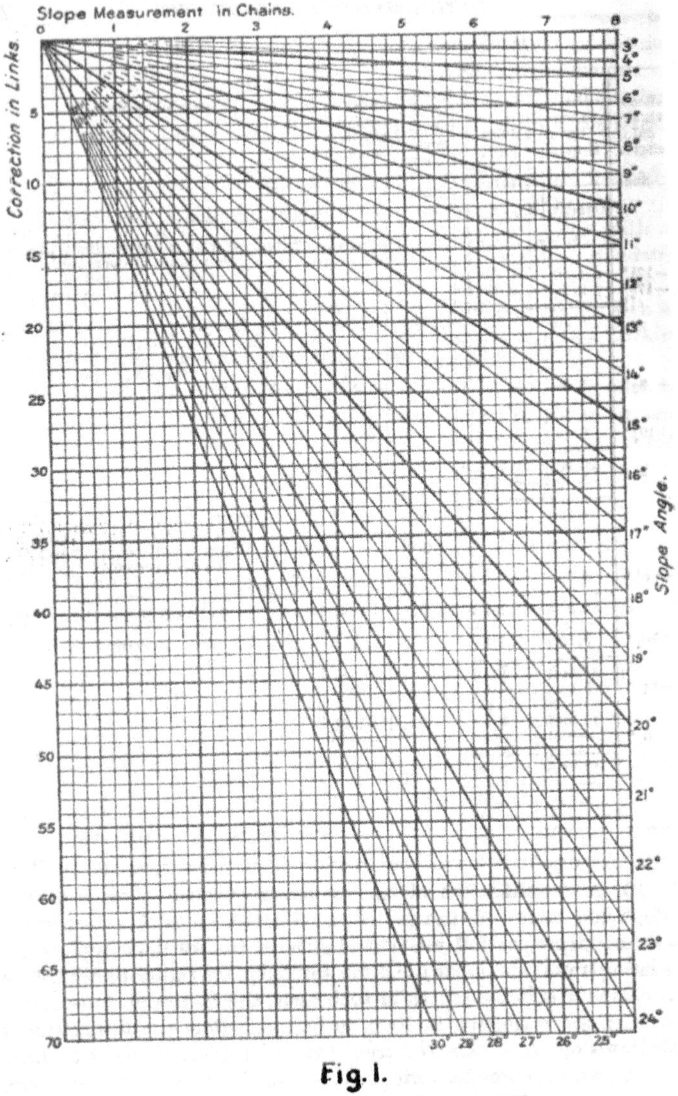


Fig. 1.

Reduction from the slope to the horizontal.

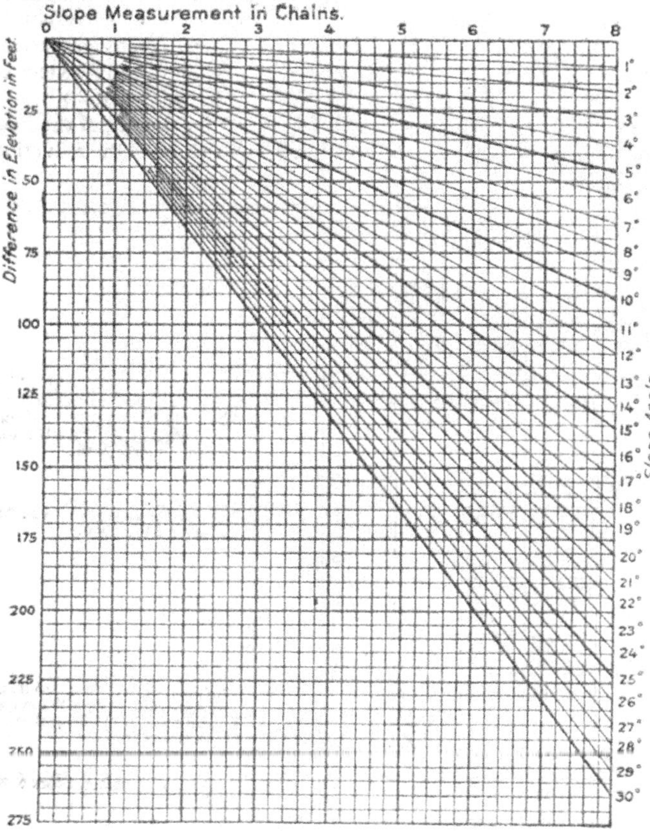


Fig. 2.

Reduction for difference of elevation.

points for the correction for the full length of the tape. The second diagram is constructed with the vertical lines representing similarly the measurement on the slope in the *chain* unit; the horizontal lines in this diagram representing the difference in elevation in *feet*, at intervals of 5 feet; slanting lines are drawn to represent various degrees of slope scaled to the proper points for the differences of elevation for the full length of the tape. (See figs. 1 and 2.)

**20.** The following is an example of record for the use of the long steel tape and clinometer, and reductions by the use of the reduction diagrams:

| Field record. | | | | | Final field notes. | |
|---|---|---|---|---|---|---|
| Mean vertical angle. | Distance on slope. | Correction to horizontal. | Intermediate measurement. | Difference in elevation. | | |
| | *Chains.* | *Chains.* | *Chains.* | *Feet.* | *Chains.* | North, bet. secs. 19 and 24. Desc. 155 ft. over NW. slope, through scattering timber and dense undergrowth. |
| −12¼° | 4.50 | 0.10 | | − 60 | | |
| −17½° | 2.20 | .10 | | − 45 | | |
| | 6.70 | 0.20 | 3.80 | − 50 | 10.30 | Dry gulch, course W.; asc. 295 ft. over SW. slope. |
| | | | | + 50 | | |
| + 8¼° | 8.00 | .08 | | + 75 | | |
| | 14.70 | 0.28 | | | | |
| +19¾° | 6.20 | .37 | | + 140 | | |
| | 20.90 | 0.65 | | | | |
| + 7¾° | 3.30 | .03 | | + 30 | | |
| | 24.20 | 0.68 | 0.00 | | 23.50 | Spur, slopes W.; desc. 185 ft. to ¼ sec. cor., over NW. slope. |
| | | | 1.20 | | 24.70 | Wagon road, bears E. and W. |
| | | | 1.90 | | 25.40 | Leave undergrowth. |
| − 6½° | 8.00 | .05 | | − 60 | | |
| | 32.20 | 0.73 | 1.15 | | 32.60 | Enter heavy timber, bears NW. and SE. |
| −10¼° | 3.70 | .06 | | − 45 | | |
| | 35.90 | 0.79 | | | | |
| −14 ° | 5.00 | .15 | | − 80 | | |
| | 40.90 | 0.94 | | | | |
| 0 ° | .04 | .00 | | | | |
| | 40.94 | 0.94 | | | | |
| | 40.00 | 0.00 | | | 40.00 | Set an iron post, etc. |

**21.** By a skillful use of the long steel tape on the slope, with correct determinations of the vertical angle, and proper reductions from the slope to the true horizontal distance, the surveyor obtains one of the most rapid and reliable methods of measurement. It is essential to make all reductions for distance as the work progresses, but the additional information regarding the amount of the ascents and descents is readily obtainable from the record at the convenience of the surveyor.

### STADIA MEASUREMENTS.

**22.** Under proper safeguards the stadia method of measurement affords a useful and reliable means of overcoming the difficulties of obtaining correct distances across water and over precipitous slopes that can not be reached with the tape. It is required that the wire interval or ratio be determined in the field by frequent tests under working conditions in comparison with steel tape measurement, solving the formula given in the Standard Field Tables (p. 221) for the value of the wire ratio with the horizontal distance known. The record of the stadia tests should be given in the field notes. It is essential to accurate stadia work that rods of approved construction be used, together with two targets and a properly adjusted rod level to secure true vertical readings; the readings at all times must be restricted to suitable atmospheric conditions and to distances permitting exact bisections of the targets. Possible criticism of the use of the stadia method is found in the failure to observe proper details and not in the reliability of the method if skillfully followed.

**23.** It is desirable to state briefly at the beginning of the field notes, with every set of returns, the general plan of making stadia measurements. The following paragraphs are illustrative of the character of such record:

"All stadia measurements are made with fixed stadia wires with a ratio of 1 : 132±, as exhibited by the tests shown in the field notes; the focal constant of the instrument is 1.2 links; the rod used is a standard Philadelphia level rod graduated to feet and equipped with two targets and a rod level; all readings are made with a vertical rod."

"All stadia measurements are made with fixed stadia wires with a ratio of 1 : 100±, as exhibited by the tests shown in the field notes; the focal constant of the instrument is 1.2 links; the rod used is a standard Troy level rod graduated to feet and equipped with two targets and a rod level; all readings are made with a vertical rod."

**24.** Notation used in stadia measurements:

Hor. dist.: The true horizontal distance from the center of the instrument to the rod.

Diff. elev.: The true vertical distance from the height of the instrument to the center point between the two targets of the rod.

"$r$": Vertical rod reading.

"$v$": Observed vertical angle.

"$K$": The wire interval or ratio.

"$c$": Distance from the center of the instrument to the object glass.

"$f$": Distance from the plane of the cross-wires to the object glass.

Hor. dist. $= K\,r\cos^2 v + (c+f)\cos v$.

Diff. elev. $= K\,r\,\frac{1}{2}\sin 2v + (c+f)\sin v$.

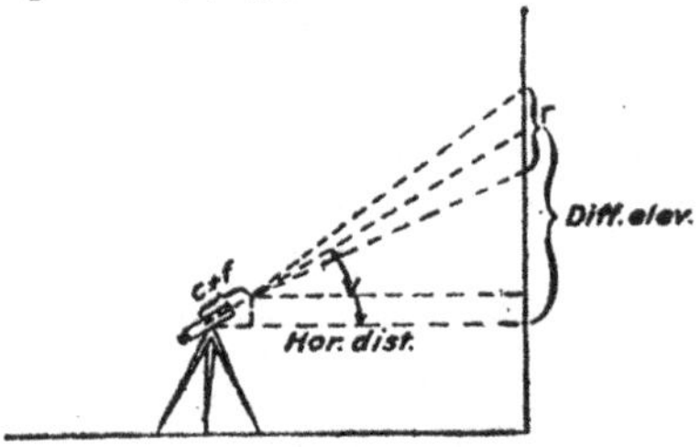


Fig. 3

**25.** In Table 6, Standard Field Tables, the natural functions "$\cos^2 v$" and "$\frac{1}{2}\sin 2v$" are tabulated by intervals of 2′ for all angles from 0° 0′ to 28° 0′; these values become natural coefficients of the rod reading in the use of the vertical rod. In the same table are tabulated the natural products "$(c+f)\cos v$" and "$(c+f)\sin v$," for three values of "$(c+f)$" which may be considered as expressed in either the link or foot unit as convenient.

**26.** In public-land surveying it is convenient to have fixed stadia wires with a ratio of 1:132, so that the sum of two rod readings in feet will be equivalent to a ratio of 1:66, or a reduced distance in chains; it is also convenient to reduce the error in the wire interval to the error in 10 chains, and to eliminate the error by applying to the reduced distance the proper correction taken from the table of proportional parts (Table 5, Standard Field Tables).

**27.** Example of test of stadia wire interval, the approximate ratio being 1:132, and the focal constant 1.2 links:

**Field record.**

| Measurement of base by steel tape and clinometer. | | | Vertical rod reading. |
|---|---|---|---|
| Mean vertical angle. | Distance on slope. | True horizontal distance. | |
| | *Chains.* | *Chains.* | *Feet.* |
| −4½° | 3.90 | 3.888 | 6.992 |
| −1¼° | 8.00 | 7.998 | 6.998 |
| +7¾° | 2.20 | 2.180 | 7.002 |
| | | | 6.995 |
| Total base | | = 14.066 | 7.003 |
| Focal constant | | = .012 | 7.004 |
| | | | 6.997 |
| Stadia base | | = 14.054 chs. | 6.995 |
| | | | 7.001 |
| " " | | = 927.564 ft. | 6.998 |
| Mean rod reading | = | | 6.9985 |
| Coefficient for 0° 40′ | = 0.9999; | | |
| 0.0001×6.9985 | = | | .0007 |
| $r \cos^2 r$ | = | | 6.9978 |

$K = \frac{927.564}{6.9978} = 132.551$

| | |
|---|---|
| Measured base | = 14.066 chs. |
| 6.9985 }<br>6.9985 } = mean rod reading. | |
| 13.997 = equivalent 1:66. | |
| 13.997 × 0.9999 = 13.996 }<br>(c+f) = .012 } | 14.008 chs. |
| Error in 14.008 chs. by stadia | = 0.058 chs. |
| Error in 10.00 chs. by stadia | = 0.041 chs. |

**Final field notes.**

June 11, 1911, I make the following test of the stadia wire interval:

| | |
|---|---|
| Horizontal length of base | = 14.066 chs. |
| Mean of 10 rod readings | = 6.9985 ft. |
| Vertical angle of of test | = −0° 40′ |
| | $K$ = 132.551 |
| Reduced error in 10 chs. | = 4.1 lks. |

All corrections to be added to the distances given by the stadia.

**28.** The error of the wire interval having been determined for a distance of 10 chains, the proportional error for any distance from 1 to 20 chains may be taken from Table 5, Standard Field Tables, thus eliminating all complex steps from the ordinary reductions of field observations.

Emphasis is placed upon the necessity for the above tests for accurate stadia work, and attention is directed to the probability that successive tests will show slightly increasing or decreasing values of the wire interval. It is not considered necessary to record in the official field notes any but the basic elements of stadia observations, omitting the details of the reductions.

**29.** The following example of record, with reductions added, is adapted to the instrument showing the above test of the wire interval, ratio 1:132 with an error of 4.1 links in 10 chains, and focal constant 1.2 links.

| Field record. | Final field notes. | |
|---|---|---|
| | *Chains.* | N. 0° 02′ W., bet. secs. 15 and 16.<br>Descend gradually over mountainous land. |
| | 12.60 | Rim of canon, bears NW. and SE.; precipitous descent of 170 ft.<br>Stadia to left bank of creek: 3.194 and 3.212 ft., −26° 44′.<br>Stadia to right bank of creek: 3.448 and 3.432 ft., −24° 10′.<br>Stadia to right rim of canon: 4.914 and 4.895 ft., +4° 58′. |
| 3.194<br>3.212<br>6.406×0.7976=5.109<br>Error + .021<br>(c+f) cos v + .011<br>12.60 +5.14 chs. | =17.74 | Left bank of creek, 62 lks. wide, course NW. |
| 6.406×0.4018=2.574 chs.<br>Diff. elev. =170 ft.<br>3.448<br>3.432<br>6.880×0.8324=5.727<br>Error + .024<br>(c+f) cos v + .011<br>12.60 +5.76 chs. | =18.36 | Right bank of creek; precipitous ascent of 225 ft. to rim of canon. |
| 5.14 chs.<br>Width of creek =0.62 chs.<br>4.914<br>4.895<br>9.809×0.9925=9.735<br>Error + .040<br>(c+f) cos v .012<br>12.60 +9.79 chs. | =22.39 | Rim of canon, bears NW. and SE.; asc. gradually. |
| 9.809×0.0863=0.847 chs.<br>= 56 ft.<br>+170 ft.<br>Diff. elev. =226 ft. | | |

**80.** Attention is directed to the fact that in making the above reductions in the chain unit, wire ratio 1:132, the process is at once resolved into taking the sum of the two rod readings in feet multiplied by the proper coefficient for vertical angle, to which product are applied the corrections for the error in the wire interval and for the horizontal value of the focal constant. As two rod readings should always be taken, one as a check upon the other, the entire operation becomes very simple. It should also be noted that in computing the difference of elevation no correction has been made for the height of the instrument above the ground, nor for the mean height of the rod reading; these corrections are compensating and ordinarily may be neglected, but in precise reductions must be considered. Therefore, in ordinary work in computing differences of elevation by the stadia method it is permissible to neglect the height of the instrument above the ground, the mean height of the rod reading, the error in the wire interval, and the term "$(c+f) \sin v$."

**31.** Many surveyors prefer the conventional stadia wire ratio 1:100 generally adopted in miscellaneous surveying practice, using a rod graduated to feet. With an instrument so fitted for public-land surveys, in which the chain unit of horizontal distance is stipulated by law, the reduction is simplified by ascertaining the logarithm of $\frac{\text{"}K\text{"}}{66}$, rod in feet and horizontal distance in chains, accomplishing the reduction of "$K r \cos^2 v$" by logarithmic functions.

**32.** Example of test of stadia wire interval, the approximate ratio being 1:100, and the focal constant 1.2 links:

| Field record.<br>Measurement of base by steel tape and clinometer.<br>Mean vertical angle. | <br><br>Distance on slope. | <br><br>True horizontal distance. | Vertical rod reading. | Final field notes. |
|---|---|---|---|---|
| | *Chains.* | *Chains.* | *Feet.* | |
| $-3\frac{3}{4}°$ | 6.40 | 6.386 chs. | 9.515 | July 7, 1915, I made the following test of the stadia wire interval:<br>Horizontal length of base = 14.160 chs.<br>Mean of ten rod readings = 9.5200 ft.<br>Vertical angle of test = + 1°51′<br>$K = 98.193$<br>log $\frac{K}{66}$, rod in feet and horizontal distance in chains = 0.172537 |
| $-4\frac{1}{2}°$ | 2.70 | 2.692 | 9.518 | |
| $+12\frac{1}{4}°$ | 5.20 | 5.082 | 9.522 | |
| | | | 9.519 | |
| Total base | | = 14.160 | 9.527 | |
| | | | 9.513 | |
| $(c+f)$ | | = .012 | 9.521 | |
| | | | 9.524 | |
| Stadia base | | = 14.148 chs. | 9.521 | |
| " " | | = 933.768 ft. | 9.520 | |
| Mean rod reading = | | | 9.5200 | |
| Coefficient for 1°54′ = 0.9989; | | | | |
| 0.0011 × 9.5200 = | | | .0105 | |
| $r \cos^2 v =$ | | | 9.5095 | |

$$K = \frac{933.768}{9.5095} = 98.193$$

log $K = 1.992081$
" $66 = 1.819544$
" $\frac{K}{66} = 0.172537$

**33.** The following example of record, with reductions added, is adapted to the instrument showing the above test of the wire interval, ratio 1:98.193 and focal constant 1.2 links.

| Field record. | Final field notes. | |
|---|---|---|
| | *Chains.* | North, bet. secs. 31 and 36.<br>Over level land. |
| | 14.20 | Commence gradual ascent of 40 ft. to base of cliff.<br>Stadia to top of cliff:<br>mean 8.472 ft., +16° 40′. |
| | 24.50 | Base of cliff, bears N. 65° W. and S. 65° E.; ascend 190 ft. to top. |
| $\log \frac{K}{66} = 0.172537$<br>" 8.472 = 0.927986<br>" $\cos^2 16° 40' = \begin{cases} 9.981361 \\ 9.981361 \end{cases}$<br>1.063245<br>nat $\frac{K}{66} r \cos^2 v = 11.568$<br>$(c+f) \cos v = .012$<br>14.20 + 11.58 chs. | 25.78 | Top of cliff; thence over level mesa. |
| $\log K = 1.992081$<br>" 8.472 = 0.927986<br>" 0.2748 = 9.439017<br>2.359084<br>Diff. elev. = 228 ft.<br>To bluff = 40<br>Cliff = 188 " | | |

**34.** Most of the General Land Office surveying instruments are equipped with fixed stadia wires of the ratio 1:132, which has been found well adapted to all practical purposes for which used, and enables the use of standard double target level rods graduated to feet. A few instruments have been provided with fixed stadia wires of the ratio 1:100, at special request, but rods graduated to links can not be furnished except upon special order, and are not purchased because they are useless except for the one purpose. Surveyors can not expect to accomplish the best results where they graduate their own rods to suit a particular instrument or personal equation.

In authorizing the use of the stadia method in the public-land surveys it is not contemplated that the same will be made a substitute for steel tape measurement where the latter is practicable, but rather that the stadia method may be used as an expedient where natural obstacles are encountered over which the distance may be more accurately measured by the stadia than otherwise, provided that every safeguard is duly observed.

### TRIANGULATIONS.

**35.** In making all triangulations for the purpose of obtaining measurements across water or over precipitous slopes, the surveyor is expected to exercise his best judgment in the selection of the measured base, and he is required to adopt the best possible geometric proportions of the sides and angles of the triangle. A complete record of the measurement of the base, the determination of the angles, the location and direction of the sides, and any other essential details of the problem will be required in the field notes, together with a small diagram to graphically represent the triangulation, but it is not considered necessary to include in the official field notes the process of the solution. The method of triangulation at all times must be sufficiently refined to produce reliable results, and when necessary to determine the value of an angle of a triangle with a precision of less than the least reading of the instrument, the method of repetitions will be employed.

**36.** In its simplest form the method of repeating an angle consists in sighting upon a station, A, with the vernier of the horizontal circle set at zero; the angle is then turned to the second station, B; the *lower clamp* is now loosened and the telescope again set upon station A with the lower tangent motion *without disturbing the angle first turned*, after which the *upper clamp* is loosened and the angle turned a second time to station B. The angle is thus "repeated" two, three, or more times, and finally the multiple angle is read, which, when divided by the repeating factor, gives a value for the angle much closer than the least reading of the instrument. For example, assume an instrument reading to single minutes of arc, and that a certain angle has been repeated five times with a resulting reading of 124° 32′; this gives a value of 24° 54′ 24″ for the angle, which if skillfully done is unquestionably closer than a single reading. In surveys which may require even greater precision both verniers are read and the angle is repeated as nearly as practicable to one complete turn of 360°, when both verniers are again read. The observer then reverses the telescope, and duplicates the process by turning the angle in the opposite direction, to eliminate instrumental errors, and finally takes a mean of the resulting four readings, which is divided by the proper factor. It is occasionally necessary in public-land surveying to repeat angles by the latter method, but the former method is of more general use and will be found dependable and quickly executed.

**37.** The base lines for triangulations are to be carefully measured, even to tenths of links if necessary, and the sum of the angles should be balanced to 180°, or redetermined if the disagreement is found to exceed 1′ of arc.

**38.** The following examples, with the reductions added, are designed to illustrate the form of record of triangulations best suited for the official field notes:

**(a) Field record.**

*Angles.*
50° 35′
93 20
36 05
180° 00′

Hor. meas. of base by one chain tape—12.80 chs.

$$\text{Dist.} = 12.80 \frac{\sin 36° 05'}{\sin 50° 35'}$$

| | |
|---|---|
| log 12.80 | =1.107210 |
| " sin 36° 05′ | =9.770087 |
| | 0.877297 |
| " sin 50° 35′ | =9.887926 |
| " 9.76 | =0.989371 |
| Dist. by tri. | =9.76 chs. |
| log hor. dist. | =0.989371 |
| " 66 | =1.819544 |
| " tan 32° 47′ | =9.808916 |
| " 415 | =2.617831 |
| Diff. elev.=415 ft. | |

**Final field notes.**

| *Chains.* | |
|---|---|
| | S. 89° 56′ W., on random line bet. secs. 19 and 30. |
| 40.00 | Set temp. ¼ sec. cor. |
| 72.20 | Top of precipitous bluff; vertical angle to flag on random line=−32° 47′; auxiliary flag bears S. 39° 21′ W.; from flag on random line the auxiliary flag bears S. 3° 16′ W., 12.80 chs. dist.; all bearings checked by direct reading of the solar, and all angles checked by deflection: |

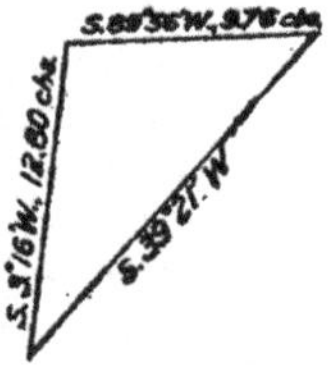


| | |
|---|---|
| Dist. on random line | =72.20 chs. |
| Dist. by triangulation | = 9.76 |
| | 81.96 |
| Dist. by return meas. | = 2.84 |
| | 79.12 |

| | |
|---|---|
| 79.12 | Intersect W. bdy. of Tp., 14 lks. S. of the cor. of secs. 19, 24, 25 and 30. |
| | Thence |
| | S. 89° 58′ E., on a true line bet. secs. 19 and 30. |
| | Ascend gradually in valley. |
| 1.75 | Base of bad-land bluff, bears N. and S.; precipitous ascent of about 400 ft. |
| 6.92 | Top of bad-land bluff, bears N. and S.; thence over level prairie. |

**(b) Field record.**

NOTE.—Stadia wire ratio, 1:132.551; $(c+f)$=1.2 lks.

| | |
|---|---|
| | 9.827 |
| | 9.839 |
| | 19.666 |
| Error= | +.082 |
| $(c+f)$= | +.012 |
| Base= | 19.76 |

*Angles.*

79° 11'
33 03
67 46

180° 00'

$$\text{Dist.} = 19.76 \frac{\sin 67° 46'}{\sin 33° 03'}$$

| | |
|---|---|
| log 19.76 | = 1.295787 |
| " sin 67° 46' | = 9.966447 |
| | 1.262234 |
| " sin 33° 03' | = 9.736692 |
| " 33.54 | = 1.525542 |
| Dist. by tri. | =33.54 chs. |

**Final field notes.**

At the meander cor. at 57.30 chs. bet. secs. 16 and 17, a flag on Indian Island bears N. 18° 41' W.; a point on a rock in the lake bears S. 82° 08' W., stadia base to this point: 9.827 and 9.839 ft., level, measured base impracticable; from point on island, flag on rock in lake bears S. 14° 22' W.; all bearings checked by direct reading of the solar, and all angles checked by deflection:

Length of base =19.76 chs.
From meander cor. to island=33.54 chs.

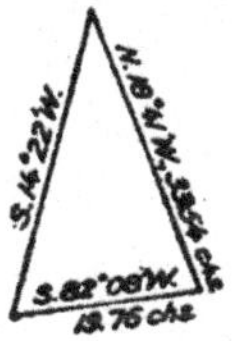


At the above point on Indian Island from which the meander cor. at 57.30 chs. bet. secs. 16 and 17, bears S. 18° 41' E., 33.54 chs. dist.; I

Set a limestone, 28 x 10 x 6 ins., 21 ins. in the ground, for auxiliary meander cor. in sec. 8, mkd. A M C on S. face; from which

A spruce, 14 ins. diam., bears N. 42½° E., 69 lks. dist., mkd. T 67 N R 43 W S 8 A M C B T.

A fir-balsam, 9 ins. diam., bears N. 14½° W., 38 lks. dist., mkd. T 67 N R 43 W S 8 A M C B T.

---

**(c) Field record.**

**Final field notes.**

| Chains. | |
|---|---|
| | 5th Guide Meridian West, through T. 14 N., between Rs. 20 and 21 W. |
| | North, bet. secs. 13 and 18. |
| | Descend 225 ft. over NW. slope, through heavy timber and dense undergrowth. |
| | Difference between measurement of 27.80 chs., by two sets of chainmen, is 4 lks.; position of middle point<br>by 1st set=27.78 chs.,<br>by 2d " =27.82 ",<br>the mean of which is

(c) **Field record, con.**

At $A = \frac{54° 29'}{3} = 18° 09' 40'' (-02'')$;

" $B = \frac{245° 13'}{3} = 81° 44' 20'' (-09'')$

" $C = \frac{240° 19'}{3} = 80° 06' 20'' (-09'')$

$180° 00' 20'' (-20'')$

$$\text{Dist.} = 16.427 \frac{\sin 80° 06' 11''}{\sin 18° 09' 38''}$$

| | |
|---|---|
| log 16.427 | =1.215558 |
| " sin 80° 06′ 11″ | =9.993488 |
| | 1.209046 |
| " " 18° 09′ 38″ | =9.493710 |
| " 51.92 | =1.715336 |
| +27.80 | |
| 79.72 | |

**Final field notes, con.**

27.80 **The south shore of Grand Lake, bears N. 62° E. and S. 48° W.**

**Set an iron post, 3 ft. long, 1 in. diam., 28 ins. in the ground, for meander cor. of frac. secs. 13 and 18, with brass cap mkd.**

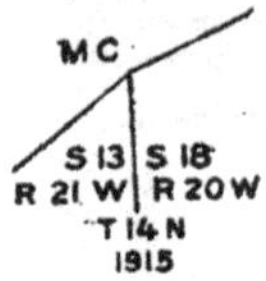


**from which**

**A pine, 8 ins. diam., bears N. 84¾° E., 105 lks. dist., mkd. T 14 N R 20 W S 18 M C B T.**

**A pine, 10 ins. diam., bears S. 26¼° W., 49 lks. dist., mkd. T 14 N R 21 W S 13 M C B T.**

**To make a triangulation across the lake I designate the above meander cor. point *A* and set a flag *B* at point for meander cor. on north shore of lake, also a flag *C* on the north shore which from point *A* bears N. 18° 09′ 38″ E.; the base *B C* bears S. 81° 44′ 11″ E., 16.427 chs. dist., the mean by two sets of chainmen,**

**by 1st set=16.425 chs.,**

**by 2d " =16.429 ",**

**longer base impracticable; the angle subtended at point *C*=80° 06′ 11″; all angles by three repetitions with error of 0′ 20″ balanced to 180°.**

**Distance across lake=51.92 chs.**

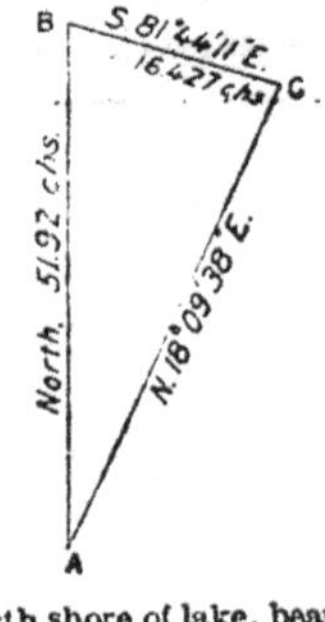


79.72 **The north shore of lake, bears S. 82° E. and N. 75° W.**

**39.** In practical field work triangulations are made only to overcome physical difficulties of measurement, and under the conditions generally presented a right-angled triangle is likely to be less desirable than an oblique triangle as the latter may be selected to fit the best topography for the base line. A stadia base may likewise be superior to a measured base as, for example, in extremely rough mountainous regions where possibly no obstruction would interfere with a good stadia determination even though a steel tape measurement of the same base might be almost impossible, or involve great delay and expense. Under some conditions a double triangulation by independent bases may be highly desirable, one result as a check upon the other, whereby the mean of the two would be a better value than either result alone. True efficiency demands a choice of the best methods to suit the peculiar conditions encountered in each circumstance, and this must be left to the judgment of the surveyor.

The subject of measurements is incomplete without a suggestion that each surveyor should devise a system of signals by means of which numbers and directions may be readily communicated from one member of a party to another; such signals will be found especially useful in long steel tape and stadia measurements and triangulations.

## INSTRUMENTS AND REQUIREMENTS AS TO THEIR ADJUSTMENT.

**40.** The direction of all lines of the public land surveys will be determined with reference to the true meridian as defined by the axis of the earth's rotation. No departure fron this rule is authorized. Beginning with the Manual of 1890 the use of the magnetic needle was prohibited except in subdividing and meandering, and then only in localities free from local attraction and with the use of suitably constructed needle instruments. The Manual of 1894 required that all surveys of the public lands of the United States, embracing all classes of lines, be made with reference to the true meridian, independently of the magnetic needle, and this prohibition against the use of the magnetic needle was even more pronounced in the Manual of 1902. In the modern instruments the length of the needle and other details relating to its construction are sacrificed in favor of the vastly more important details of design of the transit and solar attachment, and it is not presumed that the needle of the modern solar transit will give results even as reliable as those of a well-constructed needle compass. Many years' use of the solar transit and of the solar compass have proven that comparatively few localities are free from some local magnetic attraction. The needle has some value as a check and for approximate reference purposes under certain conditions, which need not be discussed in the Manual, but the use of the needle as a means of determining the direction of lines of the public-land surveys is now unqualifiedly prohibited.

**41.** Each surveyor will be supplied with one or more instruments of approved construction suited to the conditions to be encountered in his field work. It is considered desirable to include in the record of every survey, at the beginning of the first book of field notes of every set of returns, a description of the instrument used and the general method by which the azimuth determinations were accomplished. The following paragraphs suggest the form of record to be made:

"Survey commenced August 1, 1915, and executed with a Buff 'Rocky Mountain Favorite' solar transit No. 9936, 1915 model, with U-shaped standards, 4½-inch horizontal circle, 4-inch vertical circle, and improved Smith solar attachment; all azimuth determinations are accomplished with the solar attachment except the special observations upon Polaris and the sun for meridian upon which to test the solar apparatus as stated in the field notes."

"Survey commenced July 28, 1909, and executed with a Young & Sons mountain transit, No. 8070, 1907 model; the instrument is equipped with a full vertical circle and the Smith solar attachment; unless otherwise specified all azimuth determinations are accomplished with the solar attachment."

"Survey commenced May 7, 1906, and executed with a Burt solar compass made by W. & L. E. Gurley, 1905 model; unless otherwise specified all azimuth determinations are accomplished with the solar compass. The Polaris observations in camp are made with a Keuffel & Esser mountain transit No. 9699, 1903 model."

**42.** The proper supervising officer will carefully examine all instruments to see that they are in first-class condition for field work, but the burden of the final test is placed upon the surveyor who uses the instrument, as in every case the approval of an instrument will be made conditional upon satisfactory field test, the record of which will be stated in the field notes.

**43.** The record of the field test of the instrument should embrace a comprehensive statement of fact as to date, locality, and condition of the instrumental adjustments. The data relative to the independent observations for meridian should be included in the record, and the functions of apparent time, latitude and sun's declination will always be given in connection with the meridional tests of solar instruments. Various forms of record will be found in connection with the examples of observations and reductions given on the following pages.

**44.** When a transit without solar attachment is employed, Polaris observations, or direct altitude observations upon the sun, necessary to execute the work in accordance with existing law and the requirements of these instructions will be insisted upon. Observations upon Polaris, or direct altitude observations upon the sun, at frequent intervals, will be necessary to secure accuracy in the projection of transit reference lines, when solar apparatus is not used. The method of transferring the azimuth determined by the meridional observations to the surveyed lines will distinctly appear in the field notes.

**45.** Surveyors using instruments with solar apparatus will be required to make azimuth observations on Polaris, or direct altitude observations upon the sun, at the beginning of every survey, to test the accuracy of the solar apparatus, and subsequent tests will be required at least at the beginning of the subdivision of every township.

**46.** A test at the conclusion of a survey is necessary in order to prove the continued proper projection of transit lines or the continued satisfactory adjustment of the solar apparatus during the survey. A book of field notes of the survey of standard lines, or of township exteriors, will therefore show preliminary and final azimuth observations for the projection of transit lines, or preliminary and final observations and tests for the adjustment of the solar apparatus, and intermediate tests to comply with the requirements of the preceding paragraphs. The satisfactory condition of the solar apparatus at the conclusion of the subdivision of a township executed with the solar apparatus may, if so desired, be shown by specific reference to the next succeeding test preliminary to commencing the subdivision of another township included in the same series of books of subdivisional notes. A careful surveyor will make a sufficient number of tests to satisfy himself at all times of the accuracy of his alinement, but it is not intended to burden the surveyor or the field notes with superfluous evidence in this particular matter.

## GENERAL STATEMENT, TIME, LATITUDE AND AZIMUTH.

47. When considering the following treatment of field methods of determination of time, latitude and azimuth, the surveyor should bear in mind that a small error, either in assumed latitude or azimuth, produces only a slight effect in time, and when all are unknown the order of sequence in their determination should be that of time, latitude and azimuth. Time may be readily determined by the surveyor with an error not to exceed 10 seconds, while latitude and azimuth are readily determined with an error not to exceed 1′ 00″; the stated limits of error are not unreasonable where *any* of the methods herein described and authorized are employed; small errors in assumed longitude may be neglected in the determination of time, latitude and azimuth.

The following methods are limited to observations upon the sun and the north star, Polaris, and are arranged to facilitate the surveyor's work under all conditions encountered in the field, without involving more than an elementary understanding of astronomical technology. The tables and formulas published in the Standard Field Tables, and the complete daily ephemeris of the sun and Polaris and the tables of azimuths of Polaris, published in the "Ephemeris," are designed primarily for the convenience of the public-land surveyors in the field, thus encouraging a general use of approved modern methods, consistent at all times with the surveyor's clear understanding of underlying principles involved.

All reference to tables and formulas, or to the daily functions of the sun or Polaris, that follow herein, relate to the above supplements to the Manual, and when necessary to use conventional notation in the demonstrations that follow, the same agrees with that shown in detail in the Standard Field Tables.

With relation to the subject of records of observations as the same should appear in the official field notes of a survey, it must be granted that it is absolutely necessary to state all of the special basic functions of an observation, but it is quite unnecessary to include the process of reduction, except in unusual cases; thus the field notes should be complete in every respect, and it is the purpose to insist upon this requirement without involving that which is unessential to the record. In general also, no attempt is warranted by which the surveyor may endeavor to make his results by analytical reduction appear to be more accurate than justified by the refinements of the observation upon which a determination is based; but it is good practice not to discard the various small elements, fractions or decimal parts of the record value of a function until the result is ascertained, whereupon the insignificant figures may be disposed of.

### ANALYTICAL NOTATION, DECLINATION AND REFRACTION.

**48.** $\neq$: The symbol for approximation; this symbol signifies inequality, but it is used in a relation representing an inequality which approaches equality.

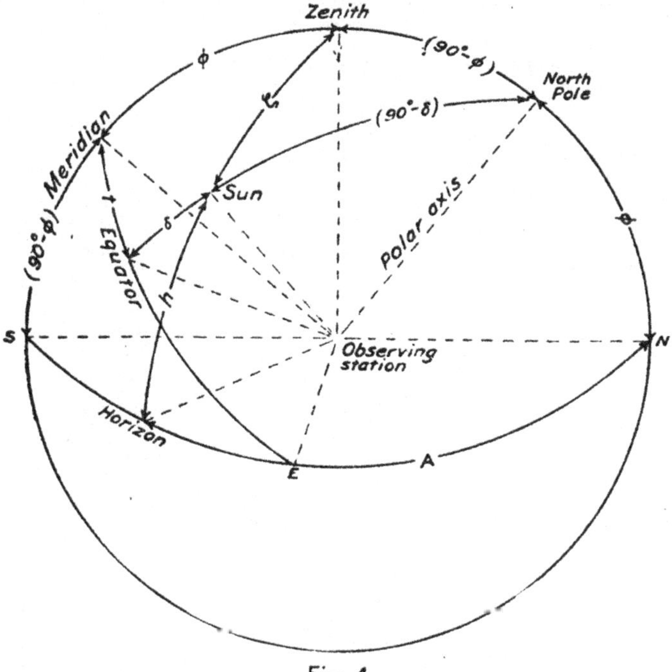


Fig. 4.

The "polo-zenith-sun" triangle as viewed from outside of the celestial sphere.

**49.** $v$: Observed vertical angle; in altitude observations on the sun, the reductions to the sun's center both vertically and horizontally, as well as instrumental errors, are eliminated by taking direct and reversed observations on the opposite limbs of the sun, and the mean observed vertical angle to the sun's center will be designated $v$ in the notation. In single observations the vertical reduction to the sun's center=16′; a refinement is had by referring to the "Ephemeris" for the value of the sun's semi-diameter for the date of observation.

**50.** *h*: True vertical angle to the sun's center, or to Polaris, in altitude observations, after correction for refraction: $h=v-$refraction in zenith distance; a refinement is had in altitude observations on the sun by adding the value of the sun's parallax=8″.9 cos $v$, opposite in effect to refraction, which results from the observer's position above the center of the earth.

**51.** $\zeta$: Zeta: true zenith distance of the sun's center:

$$\zeta=90°-h.$$

Examples of the relative use of $v$, refraction, parallax, $h$ and $\zeta$.

Field record.

| Telescope. | Watch time. | Horizontal angle. | Vertical angle. | Sun's limbs. |
|---|---|---|---|---|
| Dir. | 3h56m58s | 65° 0′ 0″ | 25° 20′ 0″ | |
| Rev. | 3 58 48 | 64 45 0 | 25 31 0 | |
| Mean | 3h57m53s | 64° 52′ 30″ | 25° 25′ 30″ | |

| | |
|---|---|
| $v$ = | 25° 25′ 30″ |
| Refraction= | − 2′ 0″ |
| Parallax= | + 0′ 8″ |
| $h$ = | 25° 23′ 38″ |
| $\zeta$ = | 64° 36′ 22″ |
| | 90° 0′ 0″ |

Final field notes.

Mar. 18, 1910, I make an altitude observation upon the sun for time and azimuth, making two observations, one each with the telescope in direct and reversed positions, observing opposite limbs of the sun.
Mean watch time of observation, 3h 57m 53s p. m.
Mean horizontal angle from flag S. to sun SW., 64° 52′ 30″.
Mean observed vertical angle 25° 25′ 30″.

Example of vertical reduction to the sun's center.

Field record.

| | |
|---|---|
| Sun's lower limb | =25° 20′ 0″ |
| Reduction to sun's center | = +16′ 6″ |
| Sun's center, $v$ | =25° 36′ 6″ |

| | |
|---|---|
| $v$ | =25° 36′ 6″ |
| Refraction | = − 2′ 0″ |
| Parallax | = + 0′ 8″ |
| $h$ | =25° 34′ 14″ |
| $\zeta$ | =64° 25′ 46″ |
| | 90° 0′ 0″ |

Final field notes.

Mar. 18, 1910, I make an altitude observation upon the sun for time, observing the sun's lower limb only; failing to observe the sun's upper limb in the reversal of the transit on account of clouds.
Watch time of observation, 3h 56m 58s p. m.
Observed vertical angle to sun's lower limb, 25° 20′ 0″, corrected to the sun's center=25° 36′ 6″.

**52.** $\phi$: Phi: Latitude of the station of observation.

**53.** $\lambda$: Lambda: Longitude of the station of observation.

**54.** $\delta$: Delta: Declination of the sun or Polaris; to be taken from the Ephemeris for the date of observation; the declination of the sun is to be corrected in hourly difference to the longitude of the station and to the time of observation; north declinations are treated as positive and south declinations as negative; a northerly hourly motion is treated as positive and a southerly hourly motion is treated as negative; in the use of the solar attachment the declination of the sun is to be corrected for refraction in polar distance, always north.

Examples of computation of the sun's declination.

(*a*) It is desired to compute the value of the sun's declination for the above altitude observation upon the sun for time and azimuth. Longitude of the station of observation, $5^h\ 8^m$ W.; apparent time of observation, $3^h\ 42^m$ p. m.:

Declination of the sun at Greenwich apparent noon
Mar. 18, 1910 $=1°\ 11'\ 3''$ S.

Difference in time from Greenwich apparent noon
to apparent time of observation:
For longitude $=\ 5^h\ 8^m$
For time, p. m. $=+3\ 42$

$8.83^h =\ 8^h\ 50^m$

Hourly difference in declination $=+59''.28$

Difference in declination from Greenwich apparent
noon to apparent time of observation:
$8.83 \times 59.28 = 523''$ $=\ 8'\ 43''$ N.

True declination of the sun $1°\ 2'\ 20''$ S.

(*b*) It is desired to prepare, by computation, a table of hourly declinations of the sun, corrected for refraction in polar distance, for use with the solar attachment, for a date March 14, 1912, and for a station in latitude $33°\ 10'$ N., and longitude $7^h\ 47^m$ W.

2° 33′ 28″.6 S. = Declination of the sun at Greenwich apparent noon, Mar. 14, 1912.

Difference in time from Greenwich apparent noon to 7 a. m., local app. time:

For longitude = $7^h\ 47^m$

For time, a. m., $12^h - 7^h\ 0^m$ = (−) 5 0

$2.78^h$ = $2^h\ 47^m$

Hourly difference in declinations = +59″.2.

2′ 44″.5 N. = Difference in declination from Greenwich apparent noon to 7 a. m., local apparent time: 2.78×59.2 = 164″.5.

2° 30′ 44″.1 S. = True declination of the sun, 7 a. m., local apparent time.

| Local apparent time. | True declination. | Refraction. | Declination setting. |
|---|---|---|---|
| 7 a. m. | 2° 30′ 44″ S. | 2′ 41″ N. | 2° 28′ 3″ S. |
| 7½ | 2 30 14 | 1 48 | 2 28 26 |
| 8 | 2 29 45 | 1 22 | 2 28 23 |
| 9 | 2 28 46 | 0 58 | 2 27 48 |
| 10 | 2 27 47 | 0 47 | 2 27 0 |
| 11 a. m. | 2 26 48 | 0 43 | 2 26 5 |
| Noon | 2 25 49 | 0 41 | 2 25 8 |
| 1 p. m. | 2 24 50 | 0 43 | 2 24 7 |
| 2 | 2 23 51 | 0 47 | 2 23 4 |
| 3 | 2 22 52 | 0 58 | 2 21 54 |
| 4 | 2 21 53 | 1 22 | 2 20 31 |
| 4½ | 2 21 23 | 1 48 | 2 19 35 |
| 5 p. m. | 2 20 54 | 2 41 | 2 18 13 |

(c) It is desired to prepare, by computation, a table of hourly declinations of the sun, corrected for refraction in polar distance, for use with the solar attachment, for a date August 12, 1912, and for a station in latitude 47° 10′ N., and longitude $7^h\ 24^m$ W.

15° 1′ 6″ N. = Declination of the sun at Greenwich apparent noon, Aug. 12, 1912.

Difference in time from Greenwich apparent noon to 6 a. m., local app. time:

For longitude = $7^h\ 24^m$

For time a.m.,

$12^h - 6^h\ 0^m$ = (−)6 0

$1.4^h$ = $1^h\ 24^m$

Hourly difference in declination = −45″.1.

1′ 3″ S. = Difference in declination from Greenwich apparent noon to 6 a. m., local apparent time: 1.4×45.1 = 63″.

15° 0′ 3″ N. = True declination of the sun, 6 a. m., local apparent time.

| Local apparent time. | True declination. | Refraction. | Declination setting. |
|---|---|---|---|
| 6 a. m. | 15° 0′ 3″ N. | 3′ 29″ N. | 15° 3′ 32″ N |
| 6½ | 14 59 40 | 2 22 | 15 2 2 |
| 7 | 14 59 18 | 1 46 | 15 1 4 |
| 8 | 14 58 33 | 1 9 | 14 59 42 |
| 9 | 14 57 48 | 0 52 | 14 58 40 |
| 10 | 14 57 3 | 0 42 | 14 57 45 |
| 11 a. m | 14 56 18 | 0 39 | 14 56 57 |
| Noon | 14 55 33 | 0 37 | 14 56 10 |
| 1 p. m. | 14 54 48 | 0 39 | 14 55 27 |
| 2 | 14 54 3 | 0 42 | 14 54 45 |
| 3 | 14 53 18 | 0 52 | 14 54 10 |
| 4 | 14 52 33 | 1 9 | 14 53 42 |
| 5 | 14 51 48 | 1 46 | 14 53 34 |
| 5½ | 14 51 26 | 2 22 | 14 53 48 |
| 6 p. m. | 14 51 3 | 3 29 | 14 54 32 |

(*d*) A graphic method for ascertaining the changing declinations of the sun, corrected for refraction in polar distance, for use with the solar attachment, is obtained by the use of a diagram constructed on cross-section paper for each date, as follows:

The horizontal lines may be used to represent each hour of the day, and the vertical lines may represent intervals of 1′ in declination. It is convenient to use the right-hand side of the sheet to represent N., and the left-hand side of the sheet to represent S., or to have N. declinations increase numerically to the right-hand side of the sheet, and S. declinations increase numerically to the left-hand side of the sheet. The vertical lines are numbered to suit the range of declination of the sun for the date. Two points are marked on the diagram to agree with the true declination of the sun; the first point is marked with the argument of declination agreeing with the declination of the sun taken from the Ephemeris for Greenwich apparent noon and with the argument of time agreeing with the local apparent time corresponding to Greenwich noon; the second point is marked agreeing with the proper declination and time 10 hours later; the straight line determined by the two points agrees with the sun's true declination for the date for the local apparent time. The proper refractions in polar distance are then scaled from the straight line to the N. for each tabulated refraction, a. m. and p. m., taken from Table 23, Standard Field Tables, appropriate to the latitude of observation and declination of the sun; the locus of the latter points is a smooth curve representing graphically the declinations of the sun, corrected for refraction in polar distance, for use with the solar attachment. The scale of the refractions must equal the scale of the intervals of 1′ in declination, and the refractions are laid off along or parallel to the horizontal lines and *not* normal to the line of true declination. At any time throughout the day the proper declination for use with the solar attachment is obtained by reference to the curve at the point corresponding to the time of observation. To obtain any true value of the sun's declination for use in the reduction of *altitude observations* reference may be made to the straight line of true declination at the point corresponding to the time of observation.

The advantage of the diagram method is found in the practical elimination of errors of computation, and the ease with which it is checked, together with the fact that in the use of the diagram actual values are obtained at any time without any process of interpolation.

The following diagrams have been prepared to illustrate the method:

DIAGRAM OF THE SUN'S DECLINATIONS.

Date, Mar. 20, 1912.
Station: Lat.=37° 30′ N.
Long.=$7^h$ $30^m$ W.

*Declination.*

Greenwich noon=0° 11′ 14″ S.=$4^h$ $30^m$ a. m.
Diff. $10^h$, +593″= 09 53 N.
0° 01′ 21″ S.=$2^h$ $30^m$ p. m.

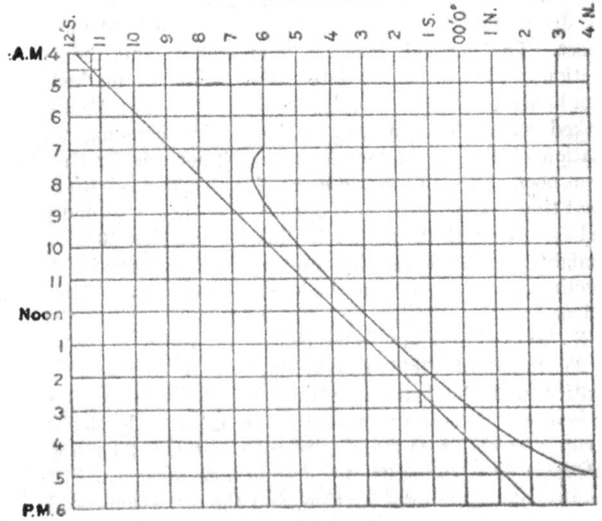


Fig. 5.

DIAGRAM OF THE SUN'S DECLINATIONS.

Date, Sept. 23, 1913.
Station: Lat. = 47° 30′ N.
Long. = $6^h$ $18^m$ W.
Greenwich noon = 0° 03′ 55″ N. = $5^h$ $42^m$ a. m.
Diff. $10^h$, −585″ = 9 45 S.
0° 05′ 50″ S. = $3^h$ $42^m$ p. m.

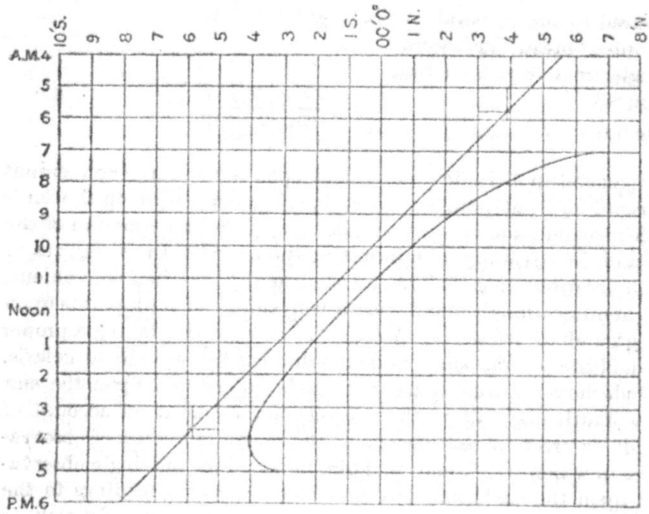


Fig. 6.

**55.** A: Azimuth angle from the true meridian to Polaris, or to the sun's center. In the following analytical examples A is referred to the north point unless otherwise noted, and the reductions are symmetrical either east or west of the meridian; all determinations for azimuth imply the recording of horizontal angles from a fixed reference point to Polaris or to the sun, or that a point has been marked on the ground to define the direction of observation; the mean horizontal angle in the first case, or the mean point in direction in the second instance, being used.

In the first of the foregoing examples of the relative use of $v$, $h$ and $\zeta$, is shown the record of certain observed horizontal angles from a fixed reference point to the sun's limbs, and now for the purpose of clearly stating the use of the notation A, the final reduction of that observation is here anticipated, in which the following result is obtained:

*Sun's azimuth.*

| | | |
|---|---|---|
| Referred to the N. point, A | =N. | 114° 07′ 28″ W. |
| Referred to the S. point, A | =S. | 65° 52′ 32″ W. |
| Recorded mean horizontal angle from flag S. to the sun SW................. | =S. | 64° 52′ 30″ W. |
| True bearing of flag......... | =S. | 1° 00′ 02″ W. |

In general in altitude observations upon the sun it is convenient to record horizontal angles from a fixed reference point to the sun's limbs; this method is preferable in view of the rapid motion of the sun and the advantage of minimizing the period of the observation. In observations upon Polaris the same method is often convenient, and at other times it may be more convenient to mark points upon the ground to define the direction of observation, taking a proper mean of the several points to define the true line of sight to Polaris.

Under adverse conditions an altitude observation upon the sun for azimuth may fail in the reversal of the transit on account of clouds or error in reading one of the angles of a series of observations, in which case it may be desirable to reduce the single observation upon the sun's limbs to equivalent corrected readings to the sun's center. In single observations on the sun, the reduction to the sun's center in azimuth $=\frac{16'}{\cos v}$; a refinement in the value of the sun's semi-idiameter is had by referring to the Ephemeris for the date of observation.

An example of reduction to the sun's center in both vertical and horizontal angles follows:

| Field record. | Final field notes. |
| --- | --- |
| Vertical angle to sun's lower limb $=25° 20' 00''$<br>Sun's semi-diameter for reduction to center $= +16' 06''$<br>Sun's center, $v$ $=25° 36' 06''$<br><br>Hor. angle from flag S. to sun's right limb, SW. $=65° 00' 00''$<br>Reduction to sun's center, $\frac{16'.1}{\cos 25° 36'}=17'.9$ $= 17' 54''$<br>Hor. angle from flag S. to sun's center, SW. $=64° 42' 06''$ | Mar. 18, 1910, I make an altitude observation upon the sun for azimuth, observing the sun's lower and right limbs only; failing to observe the sun's upper and left limbs in the reversal of the transit on account of clouds:<br>Apparent time of observation, 3h 42m p. m.<br>Observed vertical angle to sun's lower limb, 25° 20′ 00″, corrected to the sun's center = 25° 36′ 06″.<br>Observed horizontal angle to sun's right limb from flag S. to sun SW., 65° 00′ 00″, corrected to the sun's center = 64° 42′ 06″. |

**56.** Tables of mean refractions both in zenith and polar distance appear in the Standard Field Tables, arranged to meet the requirements of field use; see Tables 20 and 23. A table of coefficients to apply to mean refractions in zenith or polar distance for variations in atmospheric pressure and temperature to obtain true values of refractions is given to meet occasional necessity for its use, see Table 21. In the absence of a barometric instrument to determine the atmospheric pressure, the argument "approximate elevation above sea level" may usually be safely substituted. The differences between the true and the tabulated refractions are generally small and negligible excepting for the combined effect of low apparent altitude of observation with great elevation above sea level or extremes of temperature. The following example of reduction illustrates the method to be employed in all reductions from the tabulated refractions:

Tabulated refraction $=6' 45''=6'.75$; elevation above sea level $=10{,}000$ feet, for which elevation the coefficient is 0.70; temperature at the time of observation $=82°$ F., for which temperature the coefficient is 0.94; true refraction $=0.70\times0.94\times6'.75=4'.44=4'26''$.

## TIME.

**57.** The element of time enters into all azimuth determinations to such an extent that the surveyor should be able to arrive at the exact apparent time of all observations upon the sun and the exact local mean time of all observations upon Polaris. The sun's declination varies with the apparent time and the longitude west from

Greenwich, and enters directly into all observations upon the sun for azimuth; thus the apparent time and longitude should be known to a degree of accuracy commensurate with the refinement necessary in computing the sun's declination. The azimuth of Polaris varies with the local mean time of observation, which must be known to a degree of accuracy consistent with the result wanted in the determination of the true meridian. In observations upon Polaris at elongation precision in local mean time is unnecessary, but in hour angle observations upon Polaris it will be noted that at upper or lower culmination, in latitude 40° for example, Polaris varies 1′ in azimuth in about 2.5 minutes of time; this interval of time slowly increases toward elongation and in the latter position more than 30 minutes of time are required for a change of 1′ in azimuth.

**58.** Conversion of standard time into local mean time: watch reading ± watch error in standard time by comparison ± correction for longitude; the correction for longitude is additive east and subtractive west of the standard meridian of the time belt; the conversion table "degrees to time" (Table 18, Standard Field Tables) is convenient in this reduction.

Example of conversion of standard time into local mean time; longitude 77° 01′ 37″.5 W.:

| | |
|---|---|
| Watch time of observation | $=6^h\ 26^m\ 40^s$ p. m. |
| Watch slow of 75th meridian standard time by comparison with a standard clock | $=\ +1^m\ 22^s$ |
| Correction for longitude of station (77° 01′ 37″.5 W.$=5^h\ 08^m\ 06.5^s$) | $=\ -8^m\ 06^s$ |
| Local mean time of observation | $=6^h\ 19^m\ 56^s$ p. m. |

**59.** Conversion of apparent time into local mean time: apparent time of observation ± the equation of time; the equation of time is to be taken from the Ephemeris for the date of observation and corrected for the longitude and time of observation, conveniently interpolated as the interval from Greenwich noon to the time of observation; the watch error in local mean time is then found by taking the difference between the watch reading at the epoch of the observation and the reduced local mean time of observation.

Example of conversion of apparent time into local mean time; longitude 77° 01′ 37″.5 W.:

| | | |
|---|---|---|
| Mar. 18, 1910, apparent time of altitude observation upon sun | | $=3^h\ 42^m\ 11^s$ p. m. |
| Equation of time, Greenwich apparent noon | $+8^m\ 23.4^s$ | |
| Interpolation for longitude of station $5^h\ 08^m$ W., and time of observation $3^h\ 42^m$, p. m., $8^h\ 50^m$ after Greenwich noon, or 8.83/24 of change ($17.64^s$) in 24 hours | $= -6.5^s$ | |
| Equation of time | $+8^m\ 16.9^s$ | $+8^m\ 17^s$ |
| Local mean time of observation | | $=3^h\ 50^m\ 28^s$ |
| Watch time of observation | | $=3^h\ 57^m\ 53^s$ |
| Watch fast of local mean time | | $=\ 7^m\ 25^s$ |

### TIME IN ITS RELATION TO POLARIS OBSERVATIONS.

**60.** Polaris, a star of the second magnitude, occupies a position in the northern heavens a little more than 1° from a line defined by the axis of the earth's rotation, and on account of its brightness and proximity to the polar axis it ranks to the surveyor as the most useful circumpolar star. It will be assumed that the surveyor has learned how to identify the north star among its associates in the clear starlit heavens, especially with reference to the "pointers" in the constellation of the "Great Bear," which is popularly called the "Dipper." Polaris (α Ursæ Minoris) is nearly on a line (or great circle) determined by the pole and the star "δ Cassiopeiæ," and both stars are located in the same direction from the pole. The same line (or great circle) passes near the star "ζ Ursæ Majoris" (another star of the "Dipper"), but the latter star is located on the *opposite* side of the pole. The surveyor may note the relative position of the three stars aforementioned, if it is a clear night, and this will give an immediate indication of the approximate position of Polaris in its diurnal circle at such time of observation. The novice should secure field demonstration in these details from an experienced observer. The three stars named are all of about the same brightness. Instructions will follow (sec. 99) regarding the positive identification of Polaris by instrumental methods during the twilight period, before the star is visible to the naked eye, and the same process may

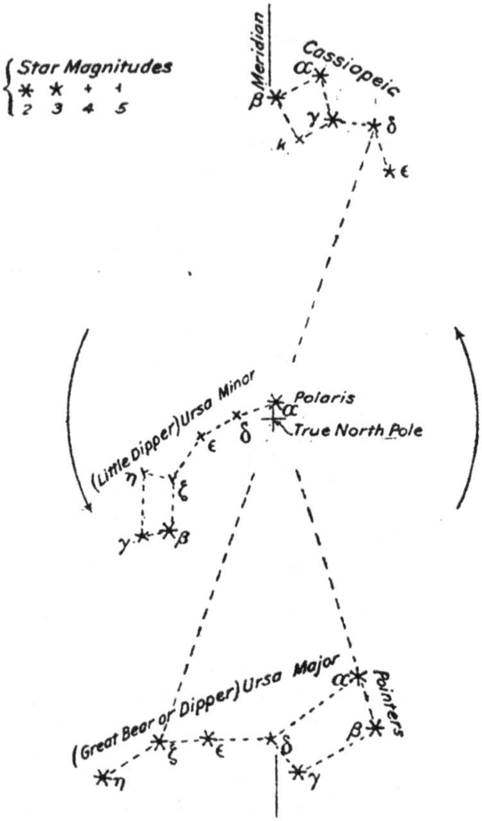


NAKED-EYE IDENTIFICATION OF POLARIS.

About noon March 23rd.
About 6 a. m. June 22nd.
About midnight September 22nd.
About 6 p. m. December 22nd.

be employed for verification of night observations, if there should be any doubt as in case the neighboring constellations are obscured by clouds.

A skillful surveyor can readily observe Polaris at sunset or sunrise without artificial illumination, and with a very clear atmosphere can perform the observation when the sun is as much as 20 or 30 minutes above the horizon. At any time that Polaris is visible any one of the various methods of Polaris observation for meridian, properly followed, is superior to any form of observation upon the sun for the same purpose. In general, in public-land surveying, the best of all practices is found in the proper use of a solar instrument adjusted to the true meridian as established by Polaris observation.

Polaris has a diurnal circle about the earth's polar axis similar to the diurnal circle of other stars, though Polaris has the smallest circle of any naked-eye star. The daily circuit of Polaris is covered in one sidereal day of 24 sidereal hours, or an equivalent of 23 hours 56 minutes 4.09 seconds of mean solar time. In its diurnal circle Polaris crosses the meridian twice, once at upper culmination, or above the polar axis, and once at lower culmination, or below the polar axis.

The direction of the apparent motion of Polaris is suggested by the following diagram:

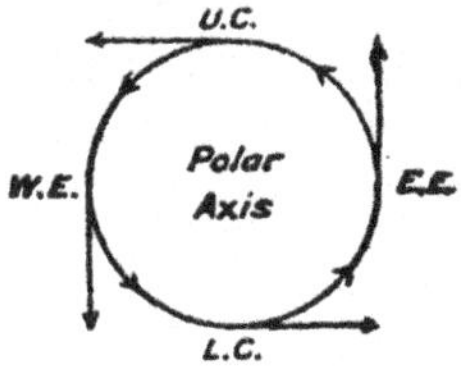


Fig. 7.

The pointings of the arrows on the above circle indicate the direction of the apparent motion of Polaris in its diurnal path, while the pointings of the arrows on the lines tangent to the circle show the direction of travel at the epochs of culmination and elongation. If the surveyor has any doubt in regard to the quadrant occupied by Polaris in its diurnal circle at the time of an observation, he may set the intersection of the telescope cross-wires exactly upon the star, then, without moving the instrument, note the direction of the star's motion and compare with the diagram.

The position of Polaris in its diurnal circle at any time may be determined by reference to the mean time interval *from* upper culmination *to* any observed position *west* of the meridian, or by reference to the mean time interval *from* any observed position *east* of the meridian *to* the succeeding upper culmination.

**61.** The Greenwich mean time of upper culmination of Polaris is tabulated in the Ephemeris for every day in the year, arranged for the ordinary civil date, a. m. or p. m.

**62.** Local mean time of upper culmination of Polaris: the Greenwich mean time of upper culmination of Polaris is to be taken from the Ephemeris for the date of observation; the amount to be subtracted from the Greenwich mean time of upper culmination of Polaris to obtain the local mean time of upper culmination, in which the argument is the longitude west from Greenwich, is obtained from the table of sidereal conversions without computation; see Table 19, Standard Field Tables.

Example of reduction from the Greenwich mean time of upper culmination of Polaris to the local mean time of upper culmination of Polaris, longitude 111° 15′ W.:

| | |
|---|---|
| Aug. 12, 1910, Gr. U. C. of Polaris | $=4^h\ 08.3^m$ a. m. |
| Red. to long. 111° 15′ W., $1^m\ 13^s$ | $= -\ 1.2$ |
| L. M. T. of U. C. of Polaris | $=4^h\ 07.1^m$ a. m. |

**63.** The Greenwich mean time of elongation of Polaris, latitude 40°, is tabulated in the Ephemeris for every day in the year, arranged for the ordinary civil date, a. m. or p. m.

**64.** Local mean time of elongation of Polaris: the mean time of elongation of Polaris, Greenwich meridian, latitude 40°, is to be taken from the Ephemeris for the date of observation: the amount to be subtracted from the mean time of elongation of Polaris, Greenwich meridian, latitude 40°, to obtain the mean time of elongation of Polaris, local meridian, latitude 40°, in which the argument is the longitude west from Greenwich, is obtained from the table of sidereal conversions (Table 19, Standard Field Tables) without computation. The amount to apply to the local mean time of elongation of Polaris latitude 40° to obtain the local mean time of elongation of Polaris at the *latitude of observation* is tabulated in the Ephemeris in connection with the table of azimuths of Polaris at elongation.

Examples of reduction from the Greenwich mean time of elongation of Polaris, latitude 40°, to the local mean time of elongation of Polaris, latitude 64° 30′ N., and longitude 146° 30′ W.:

EASTERN ELONGATION.

| | |
|---|---|
| Sept. 9, 1910, Gr. E. E. of Polaris, Lat. 40° | $=8^h\ 19.6^m$ p. m. |
| Red. to long. 146° 30′ W., $1^m\ 36^s$ | $=-\ 1.6$ |
| Red. to lat. 64° 30′ N. | $=+\ 5.8$ |
| L. M. T. of E. E. of Polaris | $8^h\ \ 23.8^m$ p. m. |

WESTERN ELONGATION, SAME STATION.

| | |
|---|---|
| Oct. 16, 1910, Gr. W.E. of Polaris, lat. 40° | $=5^h\ 48.5^m$ a. m. |
| Red. to long. 146° 30′ W., $1^m\ 36^s$ | $=\ -\ 1.6$ |
| Red. to lat. 64° 30′ N. | $=\ -\ 5.8$ |
| L. M. T. of W. E. of Polaris | $=5^h\ \ 41.1^m$ a. m. |

**65.** Conversion of a mean time interval into a sidereal time interval, or vice versa: The amount to apply to one time interval to obtain the other time interval is found in the table of sidereal conversions (Table 19, Standard Field Tables) without computation.

Example of conversion of a mean time interval into a sidereal time interval:

Mean time hour angle of Polaris for an assumed observation in Alaska = $7^h\ 32.6^m$

| | |
|---|---|
| | $=\ 7^h\ 32^m\ 36^s$ |
| Conversion into equivalent sidereal hour angle | $=\ +\ \ 1\ \ 14$ |
| Sidereal hour angle | $=\ 7^h\ 33^s\ 50''$ |

$7^h = 105°$
$33^m = \ 8°\ 15'$
$50^s = \ \ 12'\ 30''$

$=113°\ 27'\ 30''$

**66.** Hour angles of Polaris: a mean time hour angle of Polaris *west* of the meridian is the mean time interval *from* the local mean time of the last preceding upper culmination *to* the local mean time of observation of Polaris; a mean time hour angle of Polaris *east* of the meridian is the mean time interval *from* the local mean time of observation *to* the local mean time of the next succeeding upper culmination of Polaris.

The above application of the term "hour angle" is a departure from conventional usage, which has been employed in order to simplify the text. By this means one confusing step in the problem relating to hour angles for positions of Polaris east of the meridian is avoided. Polaris crosses the meridian at lower culmination at an hour angle of $11^h\ 58^m\ 02^s$, and in the arrangement of the various examples, the observations west of the meridian have been referred to the last preceding upper culmination, and those east of the meridian have been referred to the next succeeding upper culmination, thus avoiding the introduction of any hour angles exceeding $11^h\ 58^m\ 02^s$.

Examples of computing hour angles of Polaris; all taken out for longitude 117° 15′ W.:

*West of the meridian, p. m. obsn., U. C. in p. m.*

| | | |
|---|---|---|
| L. M. T. of obsn., Feb. 18, 1911 | | $=5^h\ 20.1^m$ p. m. |
| Gr. U. C. same date | $=3^h\ 36.5^m$ p. m. | |
| Red. to long. 117° 15′ W.= | − 1.3 | =3 35.2 p. m. |
| Hour angle of Polaris, west | | $=1^h\ 44.9^m$ |

*West of the meridian, p. m. obsn., U. C. in a. m.*

| | | |
|---|---|---|
| L. M. T. of obsn., May 14, 1911 | | $=\left\{\begin{array}{l}+12\\ \quad 7^h\ 12.4^m\end{array}\right.$ p. m. |
| Gr. U. C. same date | $=10^h\ 02.1^m$ a. m. | |
| Red. to long. 117° 15′ W.= | −1.3 | = 10 0.8 a. m. |
| Hour angle of Polaris, west | | $=9^h\ 11.6^m$ |

*West of the meridian, a. m. obsn., U. C. in p. m.*

| | | |
|---|---|---|
| L. M. T. of obsn., Nov. 3, 1911 | | $\begin{cases}+12 \\ = 6^h\ 31.6^m \text{ a. m.}\end{cases}$ |
| Gr. U. C., Nov. 2 | $=10^h\ 43.9^m$ p. m. | |
| Red. to long. 117° 15′ W. | −1.3 | =10 42.6 p. m. |
| Hour angle of Polaris, west | | $= 7^h\ 49.0^m$ |

*West of the meridian, a. m. obsn., U. C. in a. m.*

| | | |
|---|---|---|
| L. M. T. of obsn., Aug. 11, 1911 | | $= 5^h\ 05.9^m$ a. m. |
| Gr. U. C., same date | $=4^h\ 13.6^m$ a. m. | |
| Red. to long. 117° 15′ W. | −1.3 | = 4 12.3 a. m. |
| Hour angle of Polaris, west | | $= 0^h\ 53.6^m$ |

*East of the meridian, p. m. obsn., U. C. in p. m.*

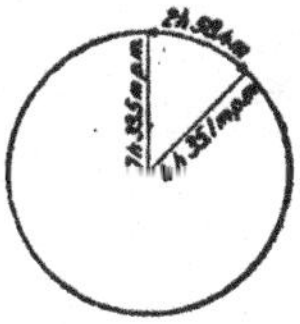

| | |
|---|---|
| Gr. U. C., Dec. 20, 1911 | $= 7^h\ 34.8^m$ p. m. |
| Red. to long. 117° 15′ W. | = −1.3 |
| L. M. T. of U. C., Dec. 20 | = 7 33.5 p. m. |
| L. M. T. of obsn., same date | = 4 35.1 p. m. |
| Hour angle of Polaris, east | $= 2^h\ 58.4^m$ |

*East of the meridian, p. m. obsn., U. C. in a. m.*

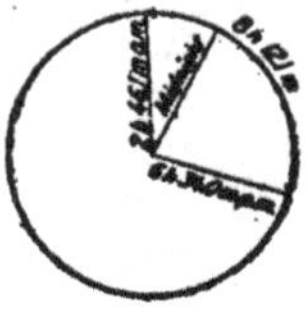

| | |
|---|---|
| Gr. U. C., Sept. 2, 1911 | $= 2^h\ 47.4^m$ a. m. |
| Red. to long. 117° 15′ W. | $= \quad -1.3$ |
| L. M. T. of U. C., Sept. 2 | $= 2\ \ 46.1$ a. m. <br> $+12$ |
| L. M. T. of obsn., Sept. 1 | $= 6\ \ 34.0$ p. m. |
| Hour angle of Polaris, east | $8^h\ 12.1^m$ |

*East of the meridian, a. m. obsn., U. C. in p. m.*

| | |
|---|---|
| Gr. U. C., Mar. 19, 1911 | $= 1^h\ 42.1^m$ p. m. |
| Red. to long. 117° 15′ W. | $= \quad -1.3$ |
| L. M. T. of U. C., Mar. 19 | $= 1\ \ 40.8$ p. m. <br> $+12$ |
| L. M. T. of obsn., same date | $= 6\ \ 06.6$ a. m. |
| Hour angle of Polaris, east | $= 7^h\ 34.2^m$ |

*East of the meridian, a. m. obsn., U. C. in a. m.*

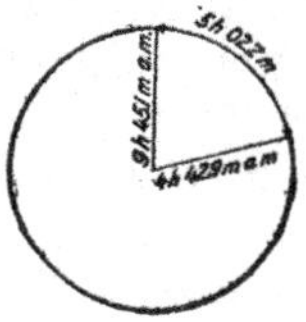


| | |
|---|---|
| Gr. U. C., May 18, 1911 | = $9^h\ 46.4^m$ a. m. |
| Red. to long. 117° 15′ W. | = −1.3 |
| L. M. T. of U. C., May 18 | = 9 45.1 a. m. |
| L. M. T. of obsn. | = 4 42.9 a. m. |
| Hour angle of Polaris, east | = $5^h\ 02.2^m$ |

**67.** By reference to the preceding diagram showing the direction of motion of Polaris in its diurnal circle, the motion at western elongation is shown to be vertically downward, and at eastern elongation the motion is shown to be vertically upward. At the epoch of either western or eastern elongation the motion of Polaris in azimuth is zero.

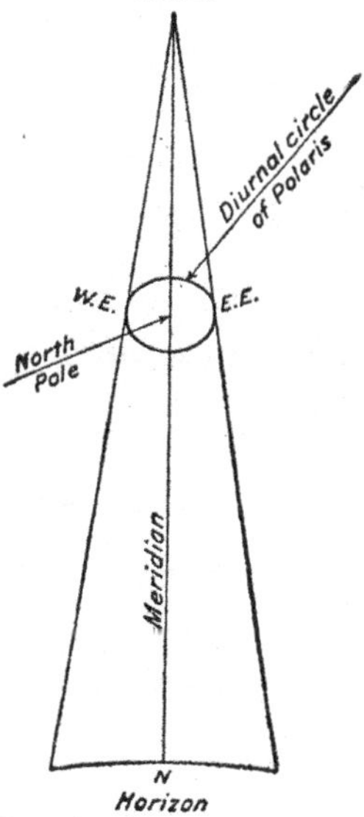


FIG. 8.—The meridian and vertical planes tangent to the diurnal circle of Polaris as viewed from inside of the celestial sphere.

At the equator, if Polaris could be observed, the hour angle of Polaris at elongation would be 90° 0′ 0″=$6^h\ 0^m\ 0^s$ sidereal hour angle=$5^h\ 59^m\ 1.02^s$ mean time hour angle, but as stations of observation are occupied in the higher latitudes the hour angle of Polaris at elongation decreases progressively. The reason for this is found in the fact that all vertical planes intersect at the zenith, and the point of tangency of a vertical plane with the diurnal circle of Polaris occurs at points corresponding to decreasing hour angles with the higher latitudes. The "spread" of the two vertical planes intersecting Polaris at eastern and western elongation increases with the higher latitudes, giving increasing azimuths at elongation with the more northern latitudes.

**68.** Mean time hour angle of Polaris at elongation: $t$=the sidereal hour angle in angular measure; this converted into time measure, and this in turn converted from a sidereal time interval into a mean time interval gives the mean time hour angle of Polaris at elongation:

$$\text{Cos } t = \text{cotan } \delta \tan \phi$$

Example of computing the mean time hour angle of Polaris at elongation, April 3, 1915, in latitude 65° 0′ N., on which date the declination of Polaris=88° 51′ 20″ N.:

| | | |
|---|---|---|
| $\phi$=65° 0′; | log tan $\phi$ | = 0.331327 |
| $\delta$=88° 51′ 20″; | " cotan $\delta$ | = 8.300530 |
| | " cos $t$ | = 8.631857 |
| Sidereal hour angle | | =87° 32′ 41″ |
| | | 87°=5h 48m |
| | | 32′= 2m 08s |
| | | 41″= 3s |
| | | =5h 50m 11s |
| Reduction to mean time hour angle | | = −0m 57s |
| Mean time hour angle at elongation | | =5h 49m 14s |

ALTITUDE OBSERVATION OF THE SUN FOR APPARENT TIME.

**69.** Altitude observation of the sun for apparent time: $t$=hour angle from apparent noon in angular measure: reverse the signs of $\delta$ for south declinations:

$$\text{Tan } \tfrac{1}{2} t = \sqrt{\frac{\sin \frac{1}{2}(\zeta+\phi-\delta) \sin \frac{1}{2}(\zeta-\phi+\delta)}{\cos \frac{1}{2}(\zeta+\phi+\delta) \cos \frac{1}{2}(\zeta-\phi-\delta)}}$$

**70.** An altitude observation of the sun for time is made by determining the correct altitude of the sun's center and recording the watch time at the epoch of observation. The following order of procedure is recommended for the elimination of instrumental errors, reduction to the sun's center, and practical elimination of differential refraction:

A. M. OBSERVATION.

Thoroughly level the transit.

Observe the sun's upper limb, recording the watch time of observation and vertical angle.

Reverse the transit.

Observe the sun's lower limb, recording the watch time of observation and vertical angle.

The mean vertical angle is equivalent to the vertical angle to the sun's center corresponding to the mean epoch of the watch readings.

###### P. M. OBSERVATION.

Thoroughly level the transit.

Observe the sun's lower limb, recording the watch time of observation and vertical angle.

Reverse the transit.

Observe the sun's upper limb, recording the watch time of observation and vertical angle.

The mean vertical angle is equivalent to the vertical angle to the sun's center corresponding to the mean epoch of the watch readings.

Example of altitude observation of the sun for apparent time:

*Final field notes.*

August 24, 1909, in latitude 37° 16′ 50″ N., and longitude 102° 12′ W., I make an altitude observation upon the sun for time, making two observations, one each with the telescope in direct and reversed positions, observing opposite limbs of the sun:

Mean observed vertical angle =19° 39′ 30″
Mean watch time of observation = $4^h\ 56^m\ 04^s$ p. m.
Watch slow of local mean time = $0^m\ 56^s$

*Field record.*

| Telescope. | Sun's limbs. | Watch time. | Vertical angle. |
|---|---|---|---|
| Direct | $\underline{\odot}$ | $4^h55^m22^s$ | 19° 33′ 00″ |
| Reversed | $\overline{\odot}$ | 4 50 40 | 19 40 00 |
| Mean | | $4^h56^m04^s$ | 19° 39′ 30″ = v |
| Refraction | | | = − 2 40 |
| Parallax | | | = + 0 08 |
| | | | h = 19° 36′ 58″ |

True vertical angle $= h = 19°37'$
Zenith distance $= \zeta = 70°23'$
Sun's declination $= \delta = 11°05'$ N.

$\zeta = 70°23'$ | $\zeta = 70°23'$
$\phi = 37\ 17$ | $\phi = 37\ 17$

$(\zeta+\phi) = 107°40'$ | $(\zeta-\phi) = 33°06'$
$\delta = 11°05'\ (+)$ | $\delta = 11°05'\ (+)$

½ values=

$(\zeta+\phi+\delta) = 118°45'$   $59°22'30''$   $22°05'30''$   $(\zeta-\phi+\delta) = 44°11'$

$(\zeta+\phi) = 107°40'$ | $(\zeta-\phi) = 33°06'$
$\delta = 11\ 05\ (+)$ | $\delta = 11\ 05\ (+)$

$(\zeta+\phi-\delta) = 96°35'$   $48°17'30''$   $11°00'30''$   $(\zeta-\phi-\delta) = 22°01'$

log $\sin \frac{1}{2}(\zeta+\phi-\delta) =$ 9.873054
" $\sin \frac{1}{2}(\zeta-\phi+\delta) =$ 9.575291

" $\cos \frac{1}{2}(\zeta+\phi+\delta) = 9.707073$   9.448345
" $\cos \frac{1}{2}(\zeta-\phi-\delta) = 9.991934$

9.699007   9.699007

" $\tan^2 \frac{1}{2}t$ = 9.749338
" $\tan \frac{1}{2}t$ = 9.874669

$\frac{1}{2}t = 36°50'42''$   $t = 73°41'24''$

$73° = 4^h52^m$
$41' = 2^m44^s$
$24'' = 2^s$

$t = 4^h54^m46^s$

Apparent time of observation $= 4^h54^m46^s$ p. m.
Equation of time $+2\ 14$

Local mean time of observation $= 4^h57^m00^s$ p. m.
Watch time of observation $= 4\ 56\ 04$ p. m.

Watch slow of local mean time $= 0^m56^s$

MERIDIAN OBSERVATION OF THE SUN FOR APPARENT NOON.

**71.** *Meridian observation of the sun for apparent noon.*—With the telescope in the meridian elevated to the sun's altitude, the watch times of transit of the sun's west and east limbs are noted, the mean of which is the watch time of apparent noon; if the observation fails for either limb the reduction to the sun's center is accomplished by adding or subtracting 68 seconds; a refinement in the amount of this time is had by referring to the Ephemeris for the time of the sun's semi-diameter passing the meridian for the date of observation; the setting for the approximate altitude of the sun's center is:

$$v \neq 90° - \phi \pm \delta$$

OBSERVING PROGRAM.

Determine the meridian by the best means at hand and compute the altitude setting for the sun.

Level the transit, place the instrument in the meridian, and elevate the telescope to the altitude of the sun's center.

Note the watch time of the sun's west limb tangent to the vertical wire.

Note the watch time of the sun's east limb tangent to the vertical wire.

Take the mean of the readings for the watch time of apparent noon from which to compute the watch error local mean time.

Example of meridian observation of the sun for apparent noon:

*Final field notes.*

August 14, 1909, in latitude 37° 16′ N., and longitude 102° 16′ W., with the telescope in the meridian and elevated to the sun's altitude, I observe the sun's transit for time, noting the watch time of transit of each limb:

Mean watch time of apparent noon = $12^h\ 00^m\ 27^s$
Watch slow of local mean time = $4^m\ 06^s$

*Field record.*

| | | |
|---|---|---|
| Setting: | | 90° 00′ |
| | $\phi \neq$ | (−) 37° 16′ N. |
| | $\delta \neq$ | (+) 14° 25′ N. |
| | $v \neq$ | 67° 09′ |

| | | |
|---|---|---|
| ɋ | Watch time of transit, W. limb | $=11^h\ 59^m\ 22^s$ |
| þ | Watch time of transit, E. limb | =12 01 32 |
| | Watch time of apparent noon | $=12^h\ 00^m\ 27^s$ |
| | Apparent noon $=12^h\ 00^m\ 00^s$ | |
| | Equation of time = + 4 33 | |
| | Local mean time of apparent noon | =12 04 33 |
| | Watch slow of local mean time | = $4^m\ 06^s$ |

The above form of meridian observation of the sun for apparent noon is by far the most convenient reliable method of time observation.

## TIME FROM THE SOLAR ATTACHMENT.

**72.** Several of the approved forms of solar apparatus, including principally the Smith solar attachment and the Burt solar compass, have a graduated arc or circle mounted normal to the polar axis to indicate the apparent time of observation. The reading of the time arc is most conveniently checked by comparison with the above form of meridian observation of the sun for apparent noon. An error in the reading of the time arc or hour circle may be corrected by adjusting the circle, or allowed for as an index error. The reading of the hour circle may then be safely accepted as giving correct apparent time for use in computing or taking out required declinations of the sun for the various forms of solar observations. A proper reading of the hour circle may be safely accepted to indicate apparent time at which moment the watch reading may be noted, and the watch error *local mean time* determined as shown in the preceding example of conversion of apparent time into local mean time; this result derived for the *watch error local mean time* may then be safely used in observations on Polaris at *elongation*, but for observations upon Polaris by the *hour angle method* the time should be determined by one of the more refined methods already given.

## LATITUDE.

**73.** It is absolutely necessary in the operation of any solar attachment to employ the correct latitude of the station, and in general in altitude observations upon the sun for azimuth or time the latitude must be well determined. In the public-land surveying practice all determinations of either time or latitude are an important part of the program of operations only so far as these functions finally enter into the establishment of the true meridian; all classes of observations given in the Manual have been arranged to facilitate the performance of solar instruments, and for this purpose a definite knowledge of the true latitude is highly important. No lack of reasonable precision should be allowed in the accepted latitude. The various forms of observations for latitude are very simple and a considerable series should be taken in every group of surveys, all reduced to the township boundaries for comparison, until a satisfactory mean has been obtained.

## MERIDIAN ALTITUDE OBSERVATION OF THE SUN FOR LATITUDE.

**74.** *Meridian altitude observation of the sun for latitude.*—Reverse the sign of $\delta$ for south declinations:

$$\phi = 90° + \delta - h$$

The following observing program is recommended:

Thoroughly level the transit and place the telescope in the meridian elevated to the sun's approximate altitude at noon.

Observe the altitude of the sun's lower limb with the sun slightly east of the meridian.

Reverse the transit.

Observe the altitude of the sun's upper limb with the sun slightly west of the meridian.

Take the mean observed vertical angle for the altitude of the sun's center at apparent noon.

The following is an example of meridian altitude observation of the sun for latitude:

*Final field notes.*

October 5, 1909, in approximate latitude 37° 20′ N., and longitude 102° 04′ W., I make a meridian altitude observation of the sun for latitude, observing the altitude of the sun's lower limb with the telescope in direct posit ion, reversing the transit and observing the sun's upper limb:

| | |
|---|---|
| Apparent time of observation, noon | $=12^h\ 00^m\ 00^s$ |
| Mean observed altitude | =47° 59′ 45″ |
| Reduced latitude | =37° 19′.3 N. |

*Field record.*

| | | |
|---|---|---|
| Setting: | | 90° 00′ |
| | $\phi \measuredangle$ (−) | 37° 20′ N. |
| | $\delta \measuredangle$ (−) | 4° 42′ S. |
| | $v \measuredangle$ | 47° 58′ |
| Lower limb | | 47° 42′ |
| Upper limb | | 48° 14′ |

| | |
|---|---|
| ☉̲ Observed alt., lower limb, tel. dir. | =47° 43′ 00″ |
| ☉̄ Observed alt., upper limb, tel. rev. | =48° 16′ 30″ |
| Mean observed altitude, $v$ | =47° 59′ 45″ |
| Refraction | − 0 52 |
| Parallax | + 0 06 |
| | $h$=47° 58′ 59″ |
| | $\delta$= 4 41 42 S. |
| $\phi$=37° 19′.3 N.=90° − $\delta$ − $h$ | =37 19 19 |
| | 90° 00′ 00″ |

**75.** The above-described observation is conveniently combined with the meridian observation of the sun for time, by observing simultaneously the sun's lower and west limbs, recording the watch time and the vertical angle and reversing the transit in the interval of about 2 minutes, and then observing simultaneously the sun's upper and east limbs. The settings for the approximate altitudes of the sun's lower and upper limbs, respectively, are:

$$v \approx 90° - \phi \pm \delta \mp 16'$$

Example of meridian observation of the sun for time and latitude:

*Final field notes.*

June 8, 1910, in approximate latitude 38° 54′ N., and longitude 77° 01′.6 W., I make a meridian observation of the sun for time and latitude, observing simultaneously the altitude of the sun's lower limb and the transit of the sun's west limb, reversing the telescope and observing simultaneously the altitude of the sun's upper limb and the transit of the sun's east limb:

| | |
|---|---|
| Mean observed altitude | =73° 55′ 30″ |
| Reduced latitude | =38° 53′.7 N. |
| Mean watch time of observation | =12$^h$ 06$^m$ 40$^s$ |
| Watch fast of local mean time | = 7$^m$ 58$^s$ |

*Field record.*

| | | |
|---|---|---|
| Setting: | | 90° 00′ |
| | $\phi \approx$ | (−) 38° 54′ N. |
| | $\delta =$ | (+) 22° 49′ N. |
| | $v \approx$ | 73° 55′ |
| Lower limb | | 73° 39′ |
| Upper limb | | 74° 11′ |

| Position of telescope. | Position of sun. | Watch time transit. | Observed vertical angle. |
|---|---|---|---|
| Direct |  | 12$^h$ 05$^m$ 37$^s$ | 73° 42′ 30″ |
| Reversed |  | 12 07 42 | 74 08 30 |
| Mean | | 12$^h$ 06$^m$ 40$^s$ | 73° 55′ 30″ |
| Refraction | | | = −16 |
| Parallax | | | = +02 |
| $h$ | | | = 73° 55′ 16″ |
| $\delta$ = 22° 49′ 00″; 90° + $\delta$ | | | = 112 49 00 |
| $\phi$ = 38° 53′.7 N. = 90° + $\delta$ − $h$ | | | = 38° 53′ 44″ |

| | | | |
|---|---|---|---|
| Watch time of apparent noon......= | $12^h$ | $06^m$ | $40^s$ |
| Apparent noon.....= | $12^h$ | $00^m$ | $00^s$ |
| Equation of time..= | | $-1$ | 18 |
| Local mean time of apparent noon.= | 11 | 58 | 42 |
| Watch fast of local mean time.....= | | $7^m$ | $58^s$ |

The known latitude of the above station is 38° 53′ 40″, but it can not be assumed that any one altitude observation of the sun will always give a result so close to the true latitude. In general a better determination of the latitude by this method is possible only by making a series of observations on successive days, or by combining the result with Polaris observations for latitude.

For the purpose of a test as to the accuracy of the above time observation, the same watch was compared with a Western Union telegraph clock as follows:

| | | | |
|---|---|---|---|
| 75th meridian time of comparison.= | $12^h$ | $00^m$ | $00^s$ |
| Correction for longitude 77° 1.6′...= | | $-08$ | 06 |
| Local mean time of comparison...= | $11^h$ | $51^m$ | $54^s$ |
| Watch time of comparison........= | 11 | 59 | 56 |
| Watch fast of local mean time.....= | | $8^m$ | $02^s$ |

ALTITUDE OBSERVATION OF POLARIS FOR LATITUDE.

**76.** ***Altitude observation of Polaris at upper culmination for latitude:***

$$\phi = h + \delta - 90^\circ$$

*Altitude observation of Polaris at lower culmination for latitude:* The mean time hour angle of Polaris at lower culmination is 11 hours 58 minutes 2 seconds:

$$\phi = h + 90^\circ - \delta$$

The settings for the approximate altitude of Polaris at upper and lower culminations, respectively, are:

$$v \neq \phi \pm (90^\circ - \delta)$$

The following program is recommended in altitude observations of Polaris at culmination for latitude.

Compute the local mean time and watch time of culmination.

Thoroughly level the transit.

About four minutes before culmination observe the altitude of Polaris with the telescope in direct position.

Reverse the transit and observe the altitude of Polaris.

Again level the transit.

Observe the altitude of Polaris with the telescope in the reversed position.

Reverse the transit to the direct position of the telescope and again observe the altitude of Polaris.

Take the mean observed altitude to use in the reduction.

Example of altitude observation of Polaris at lower culmination for latitude:

*Final field notes.*

June 19, 1910, in approximate latitude 38° 54′ N., and longitude 77° 01′.6 W., I make an altitude observation on Polaris at lower culmination for latitude, making four observations, two each with the telescope in direct and reversed positions:

| | | |
|---|---|---|
| Watch fast of 75th meridian standard time by comparison | = | $0^m\ 24^s$ |
| Mean watch time of observation | = | $7^h\ 44^m\ 37^s$ p. m. |
| Mean observed vertical angle | = | 37° 44′ 00″ |
| Reduced latitude | = | 38° 53′.4 N. |

*Field record.*

Setting: 90° 00′

$\delta \neq$ 88° 49′

$90° - \delta \neq$ 1° 11′

$\phi \neq$ 38° 54′

$v \neq 37°\ 43' = \phi - (90° - \delta)$

| | | | |
|---|---|---|---|
| Gr. U. C. of Polaris, June 19, 1910 | = | $7^h\ 39.7^m$ | a. m. |
| Reduction to longitude 77° 1.6′ W. | = | −0.8 | |
| Reduction to lower culmination | = | +11 58.0 | |
| | | $7^h\ 36.9^m$ | p. m. |
| L. M. T. of L. C. of Polaris, June 19 | = | $7^h\ 36^m\ 54^s$ | p. m. |
| Watch fast of 75th meridian standard time by comparison with a Western Union telegraph clock | = | + 0 24 | |
| Correction for longitude 77° 01′.6 W. | = | + 8 06 | |
| Computed watch time of lower culmination | = | $7^h\ 45^m\ 24^s$ | p. m. |

| Telescope. | Watch time. | Vertical angle. |
|---|---|---|
| Direct | 7h 40m 14s | 37° 42′ 30″ |
| Reversed | 7 42 45 | 37 44 30 |
| Reversed | 7 46 39 | 37 45 30 |
| Direct | 7 48 50 | 37 43 30 |
| Mean | 7h 44m 37s | 37° 44′ 00″ |
| Refraction | | −1 15 |
| | | h=37° 42′ 45″ |
| δ=88° 49′ 20″; 90°−δ | | = 1 10 40 |
| φ=38° 53′.4 N.=h+(90°−δ) | | =38° 53′ 25″ |

## AZIMUTH.

### THE SOLAR ATTACHMENT.

**77.** The solar attachment to the engineer's transit has been designed for instrumentally setting off the sides of the "pole-zenith-sun" triangle in agreement with their angular values at the station and time of observation. The sun's image may be brought into the line of collimation of an auxiliary telescope by orientation of the transit to the position where the instrumental parts are made parallel to the respective sides of the celestial triangle, whereupon the vertical plane of the "pole-zenith" arc of the solar attachment will coincide with the true meridian. Skillfully handled, the solar attachment will give at once close approximations to the true meridian comparing favorably for accuracy with direct observations. The advantage in the proper use of the solar attachment is found in its rapid and close determinations of the meridian in heavy timber, dense undergrowth, and strong wind, in low swamp or on high mountain ascents, and under nearly all other difficult physical situations encountered in the field, avoiding in its proper use accumulative errors incident to the prolongation and deflection of transit lines, and deviations in the azimuth of latitudinal lines. Several ingenious instruments have been devised for this purpose, but the Smith solar attachment, invented by Benjamin H. Smith, of Colorado, in 1880, has given the most general satisfaction of any solar instrument in meeting the special requirements of the surveying service of the General Land Office wherein it has been developed to a state of efficiency which has fully warranted the adoption of this model as a standard instrument for use in the public-land surveys.

Owing to the different details in the design of the Smith solar attachment as constructed by various instrument makers it is impossible to discuss fully the test and adjustment of each without giving a complete description of the several models, and this would lead away from the purpose of the Manual. The standard model, embracing the most recent improvements, is therefore selected for description, and discussion of the theory, adjustment and use of the Smith solar attachment. The supervising officers will furnish the surveyors with suitable instructions relative to the test and adjustment of any other special instruments supplied to them, published in circular form as deemed expedient.

DESCRIPTION.

78. The working parts of the Smith solar attachment consist of five fundamental features, each performing its own distinctive function. The principles involved have been adapted to various types of construction, and the efficiency of the different designs is related directly to the perfection which may be attained in making a proper adjustment in the field, the stability of the adjustments when made, and the compactness of the design, considering protection to the working parts and proper distribution of weight. The five fundamental working parts consist of:

1. An auxiliary telescope whose line of collimation is the polar axis of the solar attachment; the telescope may be revolved in collar bearings which are securely mounted on a vertical limb.

2. The vertical limb is mounted on a horizontal axis and has a graduated latitude arc in its vertical plane.

3. A plane mirror at the objective end of the auxiliary telescope with an axis normal to the line of collimation, and an arm leading to a graduated declination arc.

4. An hour circle on the auxiliary telescope mounted normal to the line of collimation.

5. A set of equatorial wires parallel to the axis of the reflector.

In all the forms of construction of the Smith solar attachment the auxiliary telescope is mounted in a vertical plane parallel to the transit telescope. Thus, if the instrument is in proper adjustment and oriented to the true meridian, the polar axis of the solar attachment may be made parallel to the earth's polar axis by setting off the true latitude of the station. The sun's rays are brought into the auxiliary telescope by means of the mirror, due allowance being

FIG. 9.—The solar transit as it appears in use.

made for the sun's declination north or south of the equator, but to bring the sun's image into the auxiliary telescope the latter must be revolved in its collar bearings until the reading of the hour circle agrees with the sun's apparent time. When the auxiliary telescope is thus revolved the sun's image will traverse the field of the eye-piece parallel to the equatorial wires with the limbs of the disk tangent to the same. If the transit is turned in azimuth the sun's image will immediately depart from the equatorial wires, except at noon when the image will follow the equatorial wires whether the transit be turned slightly in azimuth or the auxiliary telescope be revolved in hour angle. At apparent noon the declination arc is in a vertical plane and at this time an absolute determination may be made of the correctness of the reading of this arc.

In the modern construction the solar attachment is mounted upon the east standard of a regular light mountain model full engineer's transit, the horizontal circle of which has a diameter of 4½ inches, with a vertical circle of 4 inches diameter. The horizontal distance between the vertical planes of the transit and auxiliary telescopes is a trifle less than 4 inches. The auxiliary telescope has a focal length of 4¾ inches and a magnifying power of about 10 diameters. The latitude arc has a radius of 3 inches, and the declination arc has a radius of 3½ inches. Upon the latter arc the graduations read the true declination and, as the mirror needs to be turned only 5° to correspond to a change of 10° in the sun's declination, the graduations are made in one-half space, i. e., an interval of 10° on the arc as graduated occupies a segment of only 5°. At zero declination the plane of the mirror is at 45° to the line of sight of the auxiliary telescope. Both telescopes are fitted with the necessary colored glass shades for observing the sun. The base plate of the solar is mounted upon three foot posts, adjustable by means of opposing capstan nuts. This three-point base forms a right-angled triangle, with one side horizontal and one side vertical, thereby permitting adjustment in either of two directions: (*a*) One about a horizontal axis, and (*b*) one about a vertical axis. Suitable capstan nuts are also placed at one end of the auxiliary telescope to provide for its proper adjustment with respect to the axis of the latitude arc.

Good solar work must depend first of all upon the proper adjustment of the transit upon which it is mounted, with great care in keeping every working part cleaned, suitably oiled to work smoothly, and protected from adverse weather and injury. The same precautions are due the solar attachment. It will give very efficient meridional performance if properly adjusted and operated; nothing less can be conceded.

Before starting in with the adjustments it should be determined that the auxiliary telescope revolves smoothly in its collar bearings, neither too tight nor too loose; that there is free and smooth motion to the latitude and declination arcs; that the clamps are positive and the tangent motions smooth and free in either direction; that the eye-piece is carefully focused upon the cross wires; and that the objective is carefully focused upon any quite distant object, then secured in this position. The eye-piece turns freely and has a pin-which travels in a guide slot; this pin is not a clamp. The objective may be moved by first loosening, then pushing the screw, which will be found to travel in a guide slot near the lower (or left hand) collar bearing.

ADJUSTMENT.

**79.** The field adjustments of the solar attachment should be considered in the following order:

1. The equatorial wires must be made parallel to the axis of the reflector.

2. The line of sight of the auxiliary telescope must lie in its true turning axis.

3. The polar axis, or line of sight of the auxiliary telescope, must be normal to the axis of the latitude arc, describe a true vertical plane when turning on said axis, and said vertical plane must be parallel to the vertical plane of the transit telescope.

4. The latitude arc should read zero when the auxiliary telescope is horizontal.

5. The declination arc should at all times read the true declination of the sun plus the refraction in polar distance.

6. The hour circle should read the sun's apparent time.

There are two or more methods of testing each and every adjustment, but those stated below are without doubt the simplest, and most rapid and reliable of all field methods. The true meridian should be established by Polaris or other independent observation, upon which to test the solar, but otherwise it plays only a small part in the adjustments of the solar attachment. The true latitude of the station must be definitely known. There should be a clear view to a distant object in the horizon, but if an object less than a mile away must be utilized due allowance may be made for the horizontal distance between the vertical planes of the transit and auxiliary telescopes.

1. *The equatorial wires.*—Set up the instrument as in a regular solar observation, setting off the known latitude, declination and apparent time, and bring the sun's image accurately between the equatorial wires by orienting the transit approximately to the meridian, in which position the instrument should be clamped. (See fig. 9.) Turn the auxiliary telescope in hour angle, causing the sun's image to travel across the field from side to side. If the image follows the equatorial wires accurately the latter are parallel to the axis of the reflector as required. If the sun's image departs materially from the equatorial wires, the capstan screws which hold the diaphragm should be loosened and the reticle may be rotated until the equatorial wires are made to agree with the path of the sun's image across the field, then return each capstan screw to a proper seat.

2. *Collimation of the auxiliary telescope.*—Swing the mirror to give a direct view through the auxiliary telescope. (See fig. 10.) Set the line of sight on a distant point and clamp the instrument. Revolve the auxiliary telescope 12 hours in hour angle. If the line of sight remains fixed on the distant point it agrees with the turning axis as required. If after revolution, the line of sight appears to be above or below, or to the right or left, of the distant point, one-half of the differences should be taken up with the capstan screws which control the diaphragm. The test should be repeated until the auxiliary telescope is in perfect collimation.

3. *The polar axis.*—Carefully level the transit and then sight the main telescope to the distant point and clamp the instrument; sight toward the same point with the auxiliary telescope, and place the striding level on the latitude axis. (See fig. 10.) The striding level should be reversed to see if there is any error in the level itself, and if so take the mean position for the true indication of the level. If the latitude axis is not horizontal it may be made so by adjusting the lower pair of capstan nuts on the base frame of the solar attachment. If the line of sight of the auxiliary telescope is not parallel to that of the main telescope it may be made parallel by means of the left-hand upper pair of capstan nuts on the base frame of the solar. After fulfilling the foregoing conditions turn the transit 180° in azimuth and reverse both telescopes so as to sight again to the same distant object,

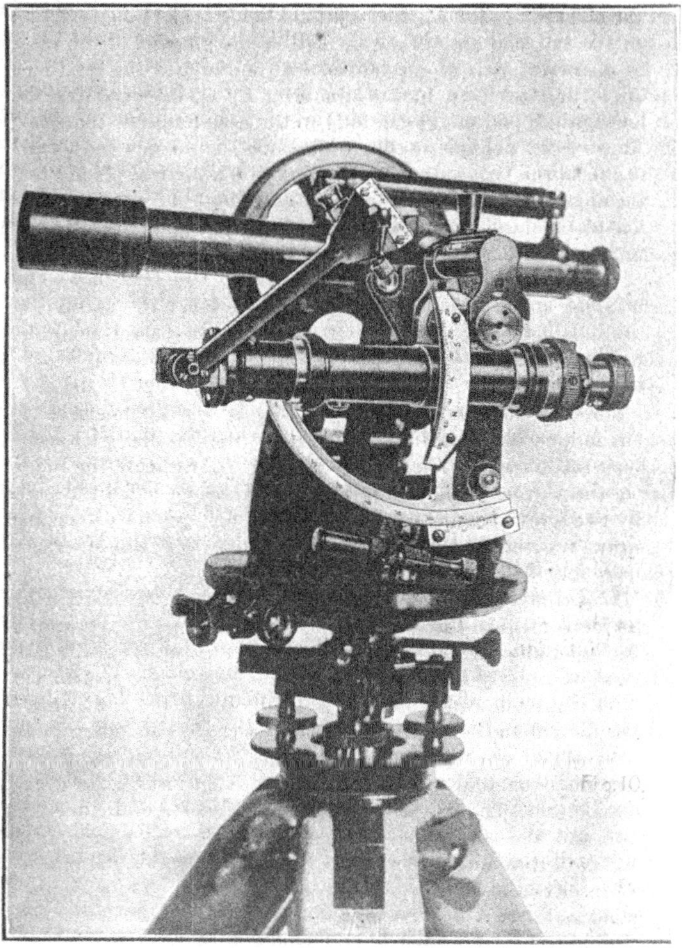

FIG. 10.—Direct sighting through the auxiliary telescope, with the mirror swung to a central position, and showing the striding level on the latitude axis.

setting the main telescope upon the object. (See fig. 11.) If the auxiliary telescope does not again sight upon the distant object, one-half the error is due to its line of sight not being at right angles to the axis of the latitude arc. Take up half of the amount of the error by means of the pair of capstan nuts at one end of the auxiliary telescope, and take up half of the error by again correcting the left-hand upper pair of capstan nuts on the base frame of the solar. The line of sight of the auxiliary telescope should now be normal to the axis of the latitude arc, should describe a vertical plane when turning on said axis, and said vertical plane should be parallel to the vertical plane of the transit telescope. The tests should be carefully repeated until the adjustments are perfected.

4. *The latitude vernier.*—Carefully level the transit, clamp the latitude arc at zero, and place the striding level in position on the auxiliary telescope. (See fig. 12.) The striding level should be reversed to see if there is any error in the level itself, and if so take the mean position for the true indication of the level. If the auxiliary telescope is not horizontal it may be made so by means of the tangent motion of the latitude arc. When the auxiliary telescope has been made truly horizontal the reading will indicate the index error of the vernier of the latitude arc. The vernier is held in position by two screws passing through elongated holes, and by loosening the screws the vernier may be shifted to read zero, or the difference from zero may be carried as an index error.

5. *The declination vernier.*—A few minutes before apparent noon set the instrument in the established meridian. Set off the known true latitude, allowing for any index error in the vernier of the latitude arc. Carefully level the transit and clamp the instrument with the main telescope in the meridian. Bring the sun's image into the field of the auxiliary telescope by turning this telescope in hour angle. At apparent noon bring the sun's image accurately between the equatorial wires by means of the tangent motion of the declination arc. The difference between the reading of the declination arc and the calculated declination (corrected for refraction) will indicate the index error of the vernier of the declination arc. This vernier is also held in position by two screws passing through elongated holes, and by loosening the screws the vernier may be shifted to read the calculated declination for apparent noon of that date, or the difference may be carried as an index error. This test should be made every day the instrument is used. If by some

FIG. 11.—The auxiliary telescope in reversed position.

failure in the adjustments of the solar attachment a difference of as much as 30″ from previous tests should be discovered in the noon observation, the new error will generally be found in one of three places: (*a*) The auxiliary telescope may be out of collimation; (*b*) the vernier of the latitude arc may have become loose and shifted; or (*c*) the vernier of the declination arc may have become loose and shifted. Any slight error in the other adjustments, or in the determination of the established meridian, will not appear in the noon test of the declination arc.

6. *The hour circle.*—A few minutes before apparent noon set the instrument in the established meridian. Level the transit and clamp the instrument with the main telescope in the meridian and elevated to the sun's altitude. Set your watch to read 12 o'clock as the sun's center crosses the vertical wire of the main telescope. At any convenient time thereafter set off the proper readings on the latitude and declination arcs, and with the instrument in the meridian, bring the sun's image to the center of the field of the auxiliary telescope and observe the watch time. If the reading of the hour circle agrees with the watch it is in adjustment; if not, it may be made to read apparent time by loosening the set screw which holds the hour circle in position and shifting the circle until the reading agrees with the watch, care being taken not to move the auxiliary telescope in hour angle until after the set screw is again seated. The test may then be repeated as often as desirable.

USE.

**80.** Before using the solar attachment the latitude of the station and the sun's declination (properly corrected for refraction in polar distance) must be known and accurately set off on the respective arcs. The instrument is carefully leveled and the apparent time set off on the hour circle. The transit is then oriented to the meridian. The plates are generally first set at zero and the sun's image brought into the field of the solar telescope before setting the lower clamp; thereupon the sun's image is brought accurately between the equatorial wires with the lower tangent motion; this gives the solar meridian. The transit may then be used for any normal function. The solar meridian may be tested as many times as may be desirable by simply setting the plates back to zero and turning the auxiliary telescope in hour angle to the apparent time; this brings the sun's image again to the center of the field. The sun's declination is constantly changing at a very slow rate, so that it is necessary

FIG. 12.—The striding level on the auxiliary telescope.

to correct the reading on the declination arc with its tangent motion to agree with the declination of the sun for the apparent time of observation.

The great advantage of the Smith solar over all other forms of solar attachment is found in the fact that the latitude and declination arcs remain clamped while the transit is being used in any normal function. Upon setting up at a second station it is necessary merely to correct the latitude and declination arcs with their tangent motions to agree with any change from the previous station. For this reason it may be operated more rapidly than any other form of solar attachment. In fact, the solar meridian is so quickly determined that the observation is usually repeated at every station.

The same restrictions which must be recognized in making direct observations on the sun operate in the same way as a prohibition in the use of any solar instrument. There are only two such limitations: (1) When the sun is within two hours, or possibly an hour and one-half of the meridian; and (2) when the sun is low in the horizon. In the first instance, the sun's relative rate of change in azimuth is much greater than the rate of change in altitude, and a small error in adjustment or in setting the arcs is greatly multiplied. In the second case the refractions are great, more or less uncertain, and changing rapidly.

The latitude of the station should always be determined with great care. Altogether too many maps are unreliable in this respect. If the latitude has been determined by competent observers, well and good, it may be free from error, but the direct altitude observation upon the sun for latitude is so simple and the reduction so easy that every operator of a solar transit should make it a practice to accomplish direct observations on the sun for latitude on as many successive days as may be necessary to give a reliable determination of the true latitude of any unknown station.

TEST.

**81.** When the solar attachment has been put in good adjustment it is proper to test it frequently on a true meridian established by Polaris observation or other approved method. The test consists merely in determining a meridian with the solar and comparing this indication with the true meridian established by other reliable method. The test should be repeated in a. m. and p. m. hours at frequent intervals, and the noon observation should most certainly be taken every day that the solar is used.

The selection of the method of observation to establish the true meridian will be made by the surveyor, the facts relative to which are to appear in the final field notes, and the solar attachment may be considered in satisfactory adjustment when all meridional tests during the usual hours of solar work are found to come within 1′ 30″ of the true meridian, whereupon the certificate of the surveyor's examination of the adjustments of his instrument will take the following form:

<table>
<tr><th>Field record.</th><th>Final field notes.</th></tr>
<tr><td>
Buff Solar Transit No. 8028.<br>
Sun's declination at Greenwich noon<br>
=5h58m a. m., local app. t. =18° 32′ 04″ S.<br>
Diff. 10h, −382″ = 6 22 S.<br>
3h58m p. m. =18° 38′ 26″ S.<br>
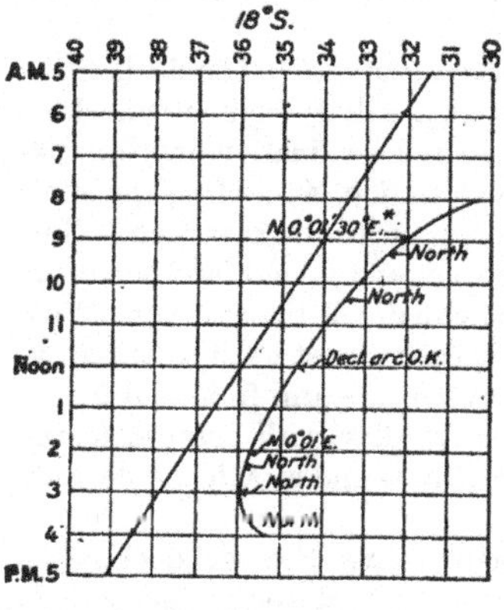

* Adjust auxiliary telescope for collimation.
</td><td>
Nov. 16, 1911, at my station in Sec. 35, T. 11 N., R. 6 E., 5th Prin. Mer., Arkansas, in latitude 35° 32.9′ N., as determined by the mean of altitude observations on the sun on Sept. 26 and 29, 1911, heretofore described, and longitude 90° 25′ W., I examine the adjustments of the instrument and correct all errors. I then test the solar apparatus by comparing its indications hourly with the true meridian established by Polaris observation Sept. 26, 1911, heretofore described.<br>
At 9h20m a. m., app. t., I set off 35° 33′ N., on the lat. arc; 18° 32′.5 S., on the decl. arc; and determine a meridian with the solar which I find to agree with the true meridian.<br>
At app. noon, with the lat. arc unchanged, I observe the sun on the meridian; the resulting reading of the decl. arc is 18° 34′.5 S., which agrees with the computed declination of the sun.<br>
At 3h0m p. m., app. t., with the lat. arc unchanged, I set off 18° 36′ S., on the decl. arc; and determine a meridian with the solar which I find to agree with the true meridian.<br>
As all of the solar observations during the usual hours of solar work come within 1′ 30″ of the true meridian, I conclude that the adjustments of the instrument are satisfactory.
</td></tr>
</table>

### THE SOLAR COMPASS.

**82.** The Burt solar compass, invented by William A. Burt, of Michigan, in 1836, was the first solar instrument, and since its introduction the instrument has been extensively used in public-land surveying; the solar compass has given general satisfaction and is still used to some extent in the public-land surveys, but in recent years it has been largely superseded by the more complete instrument already described. The Burt solar apparatus is designed for mounting upon an open-sight compass, commonly used in the early public-land surveys. A polar axis is fitted in line with the terrestrial sights when the plate verniers are set at zero. The inclination of the polar axis is controlled by a latitude arc mounted in the same vertical plane. Normal to the polar axis there is a revolving arm upon which is mounted a declination arc and two solar lines of collimation, one for north declination of the sun, and one for south declination. Each line of collimation consists of a lens and silver plate or disk mounted upon opposite ends of the revolving arm; parallel equatorial lines are drawn upon each disk symmetrical with the axis of the opposite lens. Two adjustments are peculiar to the Burt solar compass, which are here given for the surveyor's reference in the field; these adjustments should be made when the sun is within an hour of the meridian.

(1) *To make the solar lines of collimation parallel.*—The declination arm will be detached and replaced by an auxiliary frame upon which the arm will be laid. Set the latitude and declination arcs approximately correct for the hour, date and station, and bring the sun's image upon either disk as in an orientation to the meridian. Now turn the arm over, without reversing from end to end, and see if the sun's image again comes between the equatorial lines; if not, adjust the disk for half the difference and repeat the test until satisfactory. When this has been accomplished, reverse the arm from end to end for the purpose of adjusting the second disk with respect to the opposite lens. Remove the auxiliary frame and attach the declination arm in place.

(2) *To set the vernier of the declination arc.*—Set the declination vernier to read approximately zero, and bring the sun's image upon either disk as in an orientation to the meridian, changing the elevation of the polar axis as may be necessary to bring the solar line of collimation upon the sun. With the sun's image accurately between the equatorial lines, clamp all other motions and reverse the declination arm on the polar axis, thus bringing into use the second line of collimation. Note if the image of the sun is now squarely between the second pair of equatorial lines; if not, correct half the differ-

ence by movement of the tangent screw of the declination arc. Again orient in azimuth to bring the sun's image accurately between the equatorial lines, clamp and reverse as before, repeating the test until satisfactory. When the lines of collimation have thus been made truly at right angles to the polar axis, the vernier may be shifted to read zero in this position.

The general test of the Burt solar compass, by comparing its indications, resulting from solar observations made during a. m. and p. m. hours, with the true meridian determined by independent method, is similar to the test of the Smith solar attachment except in respect to the test of the latitude arc. No provision is made for independent adjustment of the latitude arc, and in the operation of the Burt solar compass the latitude is used as given by the instrument resulting from a meridian observation on the sun. In this respect therefore the noon observation with the Burt solar compass differs from the noon observation with the Smith solar attachment.

Example of noon observation with the Burt solar compass, in latitude 38° 53′ 40″ N., and longtitude 77° 01. 6′ W.:

"May 6, 1910: At this station I set off 16° 26′ N., on the decl. arc; and, at apparent noon, observe the sun on the meridian; the resulting latitude is 38° 54′ N."

## ERRORS IN AZIMUTH, DUE TO SMALL ERRORS IN DECLINATION OR LATITUDE.

**83.** It may frequently happen with a solar transit, especially at the beginning of a new survey or with an instrument insufficiently tested, that the first meridional trials are made with slight errors in the settings of the latitude and declination arcs, resulting in small errors in azimuth. This may be particularly true with a solar compass prior to a determination of the instrumental latitude. The correction of such errors has been provided for in Table 22, Standard Field Tables, which may be applied to results of single observations with considerable certainty, but not so well to a series of observations as in ordinary line work owing to the changing values (for hours from noon) of the correction coefficients. The explanation with the table gives a key to the direction of the azimuth errors on account of small errors in setting the latitude and declination arcs.

For example, at $9^{h} 40^{m}$ a. m., app. t., at a station in latitude assumed to be 46° 20′ N., a test was made with a solar transit whereby the trial indication was found to fall 0° 05′ west of the true meridian. Subsequent determinations of the true latitude of the station and of the correctness of the vernier of the declination arc showed that the actual latitude of the station was 46° 21′.5 N., and that the vernier of the declination arc had an index error which gave readings 0° 00′.5 S. of the calculated declination (i. e. reading 15° 19′.5 N. for a calculated declination of 15° 20′ N.). Thus in the test the latitude arc was set 1′.5 S. of the correct latitude of the station, and the declination arc was actually set 0′.5 N. of the value that would have been set had the index error been known.

Table 22 is entered to obtain the correction coefficients:

| Latitude. | Hours from noon. | | | |
|---|---|---|---|---|
| | 2h 0m. | 2h 20m. | 3h 0m. | |
| 45° 00′ | 2.83 | 2.55 | 2.00 | Declination coefficient. |
| 46 21.5 | ........... | 2.62 | ........... | |
| 50 00 | 3.11 | 2.81 | 2.20 | |
| 45 00 | 2.45 | 2.10 | 1.41 | Latitude coefficient. |
| 46 21.5 | ........... | 2.16 | ........... | |
| 50 00 | 2.69 | 2.31 | 1.56 | |

The corrections are then applied as follows:

| | | | |
|---|---|---|---|
| Indication of solar in test | =S. 0° | 05′.0 W. | |
| Correction for declination | = 0 | 01 .3 E. | =(2.62×0.5) |
| Correction for latitude | = 0 | 03 .2 E. | =(2.16×1.5) |
| Corrected indication of solar | =S. 0° | 00′.5 W. | |

The above corrections will often serve to explain the apparent errors of the solar, but these are not intended for use in line work, and can not be accepted in lieu of satisfactory subsequent tests based on correct values.

In the above connection it should be explained that it is not deemed desirable to burden the official record with evidence of correction for index errors found in the verniers of the latitude and declination arcs, other than to state, when such are determined, that the same are forthwith removed or are allowed for in subsequent observations.

### POLARIS AT ELONGATION.

**84.** The surveyor having thoroughly considered the theory and use of the solar instrument in its relation to the public-land surveys, and presumably mastered its operation, his attention is now directed to the approved methods of observation to establish the true meridian with which to make comparisons of the indications of the solar apparatus as a necessary test of such an instrument, or without a solar instrument, the establishment of the true meridian from which to project transit lines and to test the calculated course thereof.

Of the various independent methods of observation to establish the true meridian, the simplest and most reliable is found in the observation upon Polaris at eastern or western elongation.

Azimuth of Polaris at elongation:

$$\text{Sin } A = \frac{\cos \delta}{\cos \phi}$$

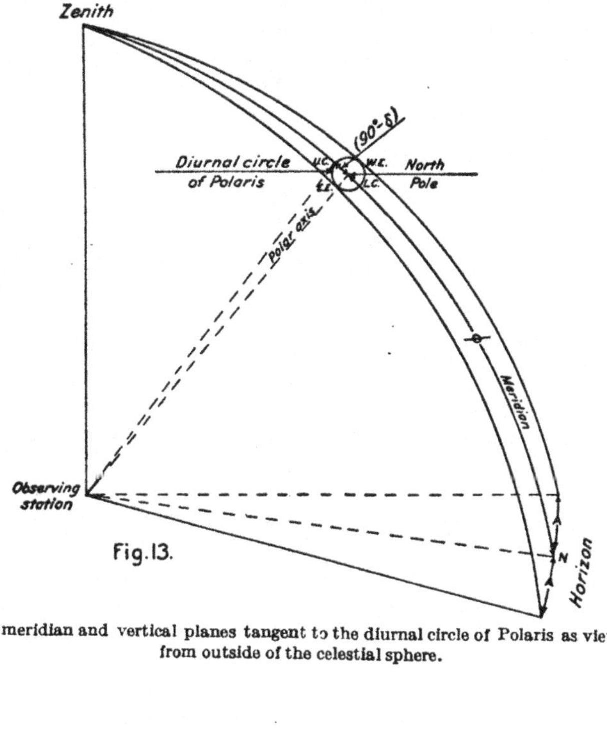


Fig. 13.

The meridian and vertical planes tangent to the diurnal circle of Polaris as viewed from outside of the celestial sphere.

Example of computing the azimuth of Polaris at elongation, October 20, 1910, in latitude 46° 20′ N., on which date the declination of Polaris=88° 49′ 48″ N.:

$$\begin{aligned} \log \cos \delta &= 8.310033 \\ \text{"} \quad \cos \phi &= 9.839140 \\ \hline \text{"} \quad \sin A &= 8.470893 \end{aligned}$$

$A$=Azimuth of Polaris at elongation=1° 41′ 41″.

**85.** A table of azimuths of Polaris at elongation for latitudes from 25° to 70° N., appears in the Ephemeris, arguments: declination of Polaris, and latitude of station.

Example in the use of the table of azimuths of Polaris at elongation, same date and station as above, showing the method of interpolation:

| Latitude. | Declination. | | |
|---|---|---|---|
| | 88° 49′ 40″ | 88° 49′ 48″ | 88° 49′ 50″. |
| | Azimuths at elongation. | | |
| 46° 00′ | 1° 41′ 15″ | 1° 41′ 04″ | 1° 41′ 01″ |
| 46 20 | | 1 41 42 | |
| 47 00 | 1 43 08 | 1 42 57 | 1 42 54 |

By interpolation in the table the required azimuth of Polaris at elongation is therefore found to be 1° 41′ 42″.

**86.** An observation upon Polaris at elongation for azimuth consists in marking upon the ground a point to define the true line of sight to Polaris at the epoch of elongation, from which to lay off the true meridian. An equivalent process is to determine the true horizontal angle by deflection from a fixed reference point to Polaris at the epoch of elongation, by which to determine the true bearing of the reference point.

POLARIS AT ELONGATION, OBSERVING PROGRAM "*a*."

**87.** Select the observing station and make suitable provision to mark the line defining the direction of Polaris at elongation; the flag point should be from 5 to 10 chains N. of the transit point, and should be cleared of all obstruction before dark. Determine the local mean and watch time of elongation of Polaris, provide suitable illumination for both the transit and flag point, and have everything in readiness as much as 15 minutes before the time of elongation.

Thoroughly level the transit.

About six minutes before elongation, with the telescope in direct position, bisect Polaris, note the watch time, and mark the direction of sight.

Reverse the transit, bisect Polaris, note the watch time, and mark the direction of sight.

Again level the transit.

With the telescope in the reverse position bisect Polaris, note the watch time, and mark the direction of sight.

Reverse the transit to the direct position of the telescope, bisect Polaris, note the watch time, and mark the direction of sight.

By daylight determine the mean (*a*) of the first and fourth sights, and (*b*) the mean of the second and third sights; then take the mean of points "*a*" and "*b*" to define the true direction of Polaris at elongation.

The mean of the four watch readings may be taken as the watch time of observation, which if within four or five minutes of correct watch time of elongation, the mean position of Polaris during the observation will be within 1″ or 2″ of true elongation. The proper value of the azimuth of Polaris at elongation having been taken from the table is then used to lay off the true meridian to the east for western elongation or to the west for eastern elongation.

The above program practically eliminates instrumental errors in observation. In laying off the azimuth of Polaris, the angle may be laid off directly, if desired, checked by the method of repetitions, and corrected if necessary; or the azimuth angle may be laid off by the natural tangent method; this should then be checked by reading the angle on the plates.

Example of observation of Polaris at elongation, observing program "*a*:"

**Field record.**

| | |
|---|---|
| Sept. 10, 1911, Gr. E. E. of Polaris, lat. 40° | = 8h 17.0m p. m. |
| Red. to long. 111° 45′ W. | = −1.2 |
| Red. to lat. 43° 22′ 30″ N. | = +0.5 |
| L. M. T. of E. E. of Polaris..... | = 8h 16.3m p. m. |
| Watch slow of l. m. t. | = −1.7 |
| Watch time of E. E. | = 8h 14.6m p. m. |

| Telescope. | Watch Time. |
|---|---|
| Direct.......................... | 8h 08m 26s p. m. |
| Reversed........................ | 8 11 20 |
| Reversed........................ | 8 14 34 |
| Direct.......................... | 8 16 46 |
| Mean.................. | 8h 12m 46s p. m. |

Declination of Polaris = 88° 49′ 54″ N.

| Latitude. | Declination. 88° 49′ 50″ | 88° 49′ 54″ | 88° 49′ 60″ |
|---|---|---|---|
| | Azimuth. | | |
| 43° 00′ 00″ | 1° 35′ 57″ | 1° 35′ 51″ | 1° 35′ 43″ |
| 43 22 30 | | 1 36 27 | |
| 44 00 00 | 1 37 33 | 1 37 27 | 1 37 19 |

**Final field notes.**

Sept. 10, 1911, in camp at the standard cor. of Tps. 1 N., Rs. 39 and 40 E., Boise Mer., in latitude 43° 22′ 30″ N., and longitude 111° 45′ W., at 8h 16.3m p. m., l. m. t., I observe Polaris at eastern elongation, making four observations, two each with the telescope in direct and reversed positions, and mark the mean point in the line thus determined, on a peg driven firmly in the ground, 5 chs. N.

Azimuth of Polaris at eastern elongation = 1° 36′ 27″.

Sept. 11: I lay off the azimuth of Polaris, 1° 36′ 30″, to the west, and mark the meridian thus determined, by a tack in a peg driven firmly in the ground, 5 chs. N.

The above program of observation of Polaris at elongation is the most convenient method where there is an opportunity to mark the direction of the line of sight. Occasionally conditions obtain where it is impossible to define or mark the direction of the observation; the program may then be altered to the reading of deflection angles as shown in the next method.

### POLARIS AT ELONGATION, OBSERVING PROGRAM "b."

88. Select the observing station and mark a point by driving a tack in a peg driven firmly in the ground approximately in the true meridian as determined by the solar before sunset, or choose other suitable reference mark in any direction. The reference point should not be nearer to the transit than 5 chains distant. Deter-

mine the local mean and watch time of elongation of Polaris, provide suitable illumination for both the transit and flag point, and have everything in readiness as much as 10 minutes before the time of elongation.

Thoroughly level the transit.

About 6 minutes before elongation with the transit in direct position, read and note the deflection angle from the reference point to Polaris, noting also the watch time of observation.

Reverse the transit and read and note the deflection angle from the reference point to Polaris, noting also the watch time of observation.

Again level the transit.

With the transit in the reverse position again read and note the deflection angle from the reference point to Polaris and note the watch time of observation.

Reverse the transit to the direct position and again read and note the deflection angle from the reference point to Polaris, and note the watch time of observation.

As the position of Polaris remains within about 0° 00′ 01″ of true elongation for a period of about five or six minutes either side of the time of exact elongation, the observation may be considered satisfactory if all of the watch readings fall within the stated period.

The mean of the four horizontal deflection angles may be taken to which must be applied the value of the azimuth of Polaris at elongation taken from the table, to obtain the true bearing of the reference flag, from which the true meridian may be laid off, or the flag may be used as a reference point.

A reference point in any direction may be used in the above method; the direction of the deflection from the reference point to Polaris should always be clearly stated. The insignificant figures of the final result may be discarded if the value of the bearing angle does not enter into another determination that demands great precision. In the example below the true meridian may be laid off by accurately measuring a distance from the reference point, at right angles to the line of sight, found by multiplying the distance from the instrument to the reference point (660 ft.) by the tangent of the bearing angle (nat tan 0° 00′ 44″=0.00021) which gives 0.14 ft. After laying off the true meridian the angle from the reference point may be checked by the method of repetitions.

Example of observation of Polaris at elongation, observing program "*b*":

**Field record.**

| | |
|---|---|
| Apr. 1, 1911, Gr. W. E. of Polaris, lat. 40° | =6h 46.1m p. m. |
| Red. to long. 104° 39′ W. | = − 1.1 |
| " " lat. 46° 13′ N. | = − 1.0 |
| L. M. T. of W. E. of Polaris. | =6h 44.0m p. m. |
| Watch slow of L. M. T. | = − 2.9 |
| Watch time of W. E | =6h 41.1m p. m. |

| Telescope. | Watch time. | Deflection angle. |
|---|---|---|
| Direct | 6h 37m 22s | 1° 41′ 00″ |
| Reversed | 6 39 40 | 1 40 30 |
| Reversed | 6 43 14 | 1 40 00 |
| Direct | 6 45 30 | 1 40 30 |
| Mean | 6h 41m 26s | 1° 40′ 30″ |

Declination of Polaris = 88° 49′ 58″ N.

| Latitude. | Declination. 88° 49′ 50″ | 88° 49′ 58″ | 88° 49′ 60″ |
|---|---|---|---|
| | Azimuth. | | |
| 46° 00′ | 1° 41′ 01″ | 1° 40′ 50″ | 1° 40′ 47″ |
| 46 13 | | 1 41 14 | |
| 47 00 | 1 42 54 | 1 42 42 | 1 42 39 |

**Final field notes.**

April 1, 1911, in camp at the cor. of Tps. 5 and 6 N., Rs. 56 and 57 E., Prin. Mer., Montana, in latitude 46° 13′ N., and longitude 104° 39′ W., at 6h 44.0m p. m., l. m. t., I observe Polaris at western elongation, making four observations, two each with the telescope in direct and reversed positions, reading the deflection angle from a tack in a peg driven firmly in the ground, 10 chs. N., west to Polaris:

| | |
|---|---|
| Azimuth of Polaris at western elongation | = 1° 41′ 14″ |
| Mean deflection angle | = 1 40 30 |
| True bearing of mark | = N. 0° 00′ 44″ W. |

89. Both of the above observing programs require the surveyor to compute in advance the correct watch time of elongation, and in so conducting the observation the minimum period is consumed in the observing program; every opportunity is also thus afforded for reversals to eliminate instrumental errors and otherwise to introduce creditable refinement. However, should the watch error be unknown, the observation may be conducted by following the motion of Polaris in azimuth during an ample period preceding elongation to insure that the epoch of the vertical motion of Polaris in its diurnal circle,

or zero motion in azimuth, is taking place, when the surveyor marks the direction of sight thus defined.

The rate of horizontal motion for the hour preceding elongation rapidly diminishes, the change in azimuth being to the west for western elongation, or to the east for eastern elongation, when Polaris will follow the vertical cross-wire, after which the motion is reversed at an accelerating rate. This suggests a third, but less refined, observing program.

### POLARIS AT ELONGATION, OBSERVING PROGRAM "C."

**90.** Select the observing station and make suitable provision to mark the line defining the direction of Polaris at elongation; provide suitable illumination for both the transit and flag point, and have everything in readiness as much as an hour before the time of elongation.

Thoroughly level the transit.

Bisect Polaris and note that the motion of the star carries it away from the vertical wire in the proper direction. As long as this motion is discernible continue the bisection of Polaris by the tangent movement. When it can not be discerned in a period of several minutes that the least lateral motion is taking place mark the direction of sight upon the ground.

Reverse and level the transit.

Again bisect Polaris and mark the direction of sight upon the ground.

Verify the position of Polaris in its diurnal circle by again bisecting the star and without changing the tangent motion note the movement of Polaris; the motion should still be nearly vertical, with a scarcely discernible movement in the opposite horizontal direction.

By daylight determine the mean of the sights, and establish the meridian by properly laying off the correct azimuth as described in observing program "*a*."

### AZIMUTH OF POLARIS AT ANY HOUR ANGLE.

**91.** While no more reliable method is at the command of the surveyor for the establishment of the true meridian than the observation upon Polaris at elongation, yet the epoch of elongation may occur at a very inconvenient time and should Polaris be obscured by clouds at the time of elongation the observation must fail. The "hour angle" method admits of observation upon Polaris for azimuth at any time that the star is visible; the precise watch error local mean time must be known, but if this has been determined, the hour angle method becomes at once the most convenient. The possible accuracy of the result compares favorably in every way with the refinement to be obtained in an observation at elongation.

The determination of the watch error local mean time and the calculation of hour angles having been fully treated on previous pages, it remains only to state that the record of the time observation should appear in the field notes with the record of all observations upon Polaris for azimuth by the hour angle method, as the azimuth observation is incomplete without the time determination. With the meridian observation of the sun for apparent noon, and the use of the azimuth tables contained in the Ephemeris, the entire process becomes so simple and yet so highly refined that the surveyor should early become thoroughly familiar with the hour angle method.

**92.** *Azimuth of Polaris at any hour angle.*—"$t$"=sidereal hour angle in angular measure; in hour angles exceeding 90° the function "$-\sin\phi\cos t$" becomes positive by virtue of the cosine of an angle between 90° and 270° being treated as negative in analytical reductions:

$$\text{Tan } A=\frac{\sin t}{\cos\phi\tan\delta-\sin\phi\cos t}$$

Example of computing the azimuth of Polaris, February 23, 1911, at a mean time hour angle of $2^h\ 37.4^m$, in latitude 33° 20′ N., on which date the declination of Polaris=88° 50′ 08″ N.:

| | | | |
|---|---|---|---|
| Mean time hour angle | $=2^h37.4^m$ | | |
| | $=2^h37^m24^s$ | $2^h$ | $=30°$ |
| | | $37^m$ | $=\ 9°15'$ |
| Red. to sidereal hour angle= | $+26^s$ | $50^s$ | $=\ \ 12'\ 30''$ |
| Sidereal hour angle | $=2^h37^m50^s$ | | $=39°27'30''$ |

| | | | |
|---|---|---|---|
| $\log\cos\phi$ | $=9.921940$ | $\log\sin\phi$ | $=9.739975$ |
| " $\tan\delta$ | $=1.691944$ | " $\cos t$ | $=9.887666$ |
| " $\cos\phi\tan\delta$ | $=1.613884$ | " $\sin\phi\cos t$ | $=9.627641$ |
| nat $\cos\phi\tan\delta$ | $=41.104$ | nat $\sin\phi\cos t$ | $=0.424$ |
| nat $\sin\phi\cos t$ | $=0.424\ (-)$ | | |
| | | $\log\sin t$ | $=9.803127$ |
| Algebraic sum | $=40.680$ | " 40.680 | $=1.609381$ |
| | | " $\tan A$ | $=8.193746$ |

Azimuth of Polaris at above hour angle, $A=0°\ 53'\ 42''$

**93.** A table of azimuths of Polaris at all hour angles, for latitudes from 30° to 50° N., appears in the Ephemeris, arguments: declination of Polaris, mean time hour angle, and latitude of station. For other than the latitudes given in the table the surveyor will be required to solve the above equation.

Example in the use of the table of azimuths of Polaris at any hour angle, same date, hour angle and station as above, showing the method of interpolation:

| Declination. | | | Latitude. | | |
|---|---|---|---|---|---|
| 88° 50′ 0″ | 88° 50′ 08″ | 88° 50′ 10″ | 32° 00′ | 33° 20′ | 34° 00′ |
| Mean time hour angles. | | | Azimuths of Polaris. | | |
| $2^h$ 34.7$^m$ | $2^h$ 35.1$^m$ | $2^h$ 35.2$^m$ | 52′.2 | 53′.1 | 53′.5 |
| | 37.4 | | | 53.7 | |
| 41.1 | 41.5 | 41.6 | 54.0 | 54.9 | 55.3 |

By interpolation in the table the required azimuth of Polaris is therefore found to be 0° 53′.7=0° 53′ 42″.

**94.** Example of computing the azimuth of Polaris, Sept. 11, 1911, at a mean time hour angle of $7^h$ 25.1$^m$, in latitude 42° 54′ N., on which date the declination of Polaris=88° 49′ 54″ N.:

| | | | |
|---|---|---|---|
| Mean time hour angle.............. | $=7^h\ 25.1^m$ | $7^h$ | =105° |
| | $=7^h\ 25^m\ 06^s$ | $26^m$= | 6°30′ |
| Reduction to sidereal hour angle.... | $= +1^m\ 13^s$ | $19^s$ = | 4′ 45″ |
| Sidereal hour angle................ | $=7^h\ 26^m\ 19^s$ | | =111° 34′ 45″ |

| | | | |
|---|---|---|---|
| log cos $\phi$ | =9.864833 | log sin $\phi$ | =9.832969 |
| " tan $\delta$ | =1.690496 | " cos $t$ | =9.565596 |
| " cos $\phi$ tan $\delta$ | =1.555329 | " sin $\phi$ cos $t$ | =9.398565 |
| nat cos $\phi$ tan $\delta$ | =35.919 | nat sin $\phi$ cos $t$ | = .250 |
| " sin $\phi$ cos $\delta$ | = .250(+) | log sin $t$ | =9.968441 |
| Algebraic sum | =36.169 | " 36.169 | =1.558337 |
| | | " tan $A$ | =8.410104 |
| Azimuth of Polaris at above hour angle, A | | | =1°28′22″ |

**95.** Example in the use of the table of azimuths of Polaris at any hour angle, same date, hour angle and station as above:

| Declination. | | | Latitude. | | |
|---|---|---|---|---|---|
| 88° 49′ 50″ | 88° 49′ 54″ | 88° 49′ 60″ | 42° 00′ | 42° 54′ | 44° 00′ |
| Mean time hour angles. | | | Azimuths of Polaris. | | |
| $7^h\ 15.7^m$ | $7^h\ 15.1^m$<br>25.1 | $7^h\ 14.2^m$ | 88′.6 | 89′.9<br>88. 4 | 91′.5 |
| 29.3 | 28.8 | 28.0 | 86. 6 | 87. 8 | 89. 4 |

By interpolation in the table the required azimuth of Polaris is therefore found to be 88′.4=1° 28′ 24″.

**96.** An observation upon Polaris for azimuth by the hour angle method consists in marking upon the ground a point to define the true line of sight to Polaris at any convenient epoch, the watch error local mean time being known, from which line to lay off the true meridian. An equivalent process is to determine the true horizontal angle by deflection from a fixed reference point to Polaris at any convenient epoch, the watch error local mean time being known, by which to determine the true bearing of the reference point.

HOUR ANGLE OBSERVATION OF POLARIS, OBSERVING PROGRAM "*a*."

**97.** Select the observing station and make suitable provision to mark the line defining the direction of Polaris; the flag point should be from 5 to 10 chains north of the transit point; provide suitable illumination for both the transit and flag point.

Thoroughly level the transit.

With the telescope in the direct position, bisect Polaris, note the watch time, and mark the direction of sight.

Reverse the transit, bisect Polaris, note the watch time, and mark the direction of sight.

Again level the transit.

With the telescope in the reverse position bisect Polaris, note the watch time, and mark the direction of sight.

Reverse the transit to the direct position of the telescope, bisect Polaris, note the watch time, and mark the direction of sight.

By daylight determine the mean (*a*) of the first and fourth sights, and (*b*) of the second and third sights; then take the mean of points "*a*" and "*b*" to define the true direction of Polaris at the epoch of the average of the watch times of observation.

Treat the reduction as one observation, applying the watch error to the average watch time of observation to obtain the correct local mean time of observation.

Enter the table in the Ephemeris or make the computation to determine the value of the azimuth of Polaris at the epoch of the observation with the stated arguments: declination of Polaris, mean time hour angle and latitude; this value is then used to lay off the true meridian to the east if Polaris is observed west of the meridian or to the west if Polaris is observed east of the meridian.

Example of hour angle observation of Polaris, observing program "*a*":

**Field record.**

Meridian observation of the sun for apparent noon:

| | |
|---|---|
| $\phi = 37° 18' N.$ | $90° 00'$ |
| $\delta = 4\ 36\ S.$ | $41\ 54$ |
| $\phi + \delta = 41° 54'$ | $v = 48° 06'$ |
| | *Watch time.* |
| Sun's W. limb | $= 12^h 00^m 18^s$ |
| " E. " | $= 12\ 02\ 26$ |
| Watch time of app. noon | $= 12^h 01^m 22^s$ |
| App. noon $= 12^h 00^m 00^s$ | |
| Equation of time $= -11\ 25$ | |
| L. M. T. of apparent noon | $= 11\ 48\ 35$ |
| Watch fast of l. m. t. | $12^m 47^s$ |

Hour angle observation of Polaris:

| Telescope. | Watch time. |
|---|---|
| Direct | $5^h 48^m 40^s$ p. m. |
| Reversed | 5 49 49 |
| Reversed | 5 51 36 |
| Direct | 5 52 54 |
| Mean | $5^h 50^m 45^s$ p. m. |
| Watch fast of l. m. t. | − 12 47 |
| L. M. T. of obsn | $5^h 37^m 58^s$ p. m. |
| | $5^h 38.0^m$ p. m. |

**Final field notes.**

Oct. 5, 1910, in camp at the cor. of secs. 5, 6, 31, and 32, on the S. bdy. of T. 31 S., R. 42 W., 6th Prin. Mer., Colo., in latitude 37° 17′.6 N., and longitude 102° 11′ W., I make a meridian observation of the sun for apparent noon:

Watch time of obsn. $= 12^h 01^m 22^s$.
Watch fast of l. m. t. $= 12^m 47^s$.

At the same station, at $5^h 38.0^m$ p. m., l. m. t., I make an hour angle observation on Polaris east of the meridian, making four observations, two each with the telescope in direct and reversed positions, and mark the mean point in the line thus determined, on a peg driven firmly in the ground, 8 chs. N.

| Field record, con. | Final field notes, con. |
|---|---|
| Gr. U. C. of Polaris, Oct. 6, 1910 = 0h 32.7m a. m.<br>Red. to long. 102° 11′ W. = − 1.1<br>L. M. T., U. C. of Polaris, Oct. 6 = 0h 31.6m a. m.<br>+12<br>L. M. T. of obsn., Oct. 5 = 5 38.0 p. m.<br>Hour angle of Polaris east of the meridian = 6h 53.6m<br>Declination of Polaris = 88° 49′ 42″ N. | Watch time of obsn., mean of four readings = 5h 50m 45s p. m. |

| Declination. | | | Latitude. | | |
|---|---|---|---|---|---|
| +88° 49′ | | | 36° 00′ | 37° 18′ | 38° 00′ |
| 40″ | 42″ | 50″ | | | |
| Mean time hour angles. | | | Azimuth of Polaris. | | |
| 6h 51.6m | 6h 51.2m | 6h 49.5m | 84′.3 | 85′.7 | 86′.5 |
| | 53.6 | | | 85.5 | |
| 61.5 | 61.1 | 59.6 | 83.3 | 84.8 | 85.6 |

Oct. 6, I lay off the azimuth of Polaris, 1° 25′ 30″, to the west, and mark the meridian thus determined, by a tack in a peg driven firmly in the ground, 8 chs. N.

HOUR ANGLE OBSERVATION OF POLARIS, OBSERVING PROGRAM "b."

**98.** Select the observing station and choose a suitable reference mark in any direction. The reference point should be at least 5 chains distant.

Thoroughly level the transit.

With the telescope in the direct position, read and note the horizontal angle from the reference point to Polaris, noting the watch time at the moment Polaris is properly bisected.

Reverse the transit and read and note the horizontal angle from the reference point to Polaris, noting the watch time at the moment Polaris is properly bisected.

Again level the transit.

With the telescope in the reverse position again read and note the horizontal angle from the reference point to Polaris, noting the watch time at the moment Polaris is properly bisected.

Reverse the transit to the direct position of the telescope and again read and note the horizontal angle from the reference point to Polaris, noting the watch time at the moment Polaris is properly bisected.

Treat the reduction as one observation, applying the watch error to the average watch time of observation to obtain the correct local mean time of observation.

**The mean of the four horizontal deflection angles may be taken, to which must be applied the proper value of the azimuth of Polaris at the mean epoch of the observation, to give the true bearing of the reference flag, from which the true meridian may be laid off, or the flag may be used for a reference point.**

**Example of hour angle observation of Polaris, observing program "*b*":**

### Field record.

Hour angle observation on Polaris:

| Telescope. | Horizontal angle from flag to Polaris. | Watch time. |
|---|---|---|
| Direct.......... | 177° 34′ 30″ | $6^h$ $22^m$ $37^s$ p. m. |
| Reversed...... | 177 34 30 | 6 25 29 |
| Reversed...... | 177 34 00 | 6 28 17 |
| Direct.......... | 177 34 00 | 6 30 17 |
| Mean... | 177° 34′ 15″ | $6^h$ $26^m$ $40^s$ p. m. |

| | |
|---|---|
| Watch slow of 75th mer. standard time | + 1 22 |
| Correction for longitude.......... | − 8 6 |
| L. M. T. of obsn. Mar. 21, 1910= | $6^h$ $19^m$ $56^s$ |
| | $6^h$ 19. $9^m$ p. m. |
| Gr. U. C. of Polaris, same date.....=$1^h$ 33. $0^m$ p. m. | |
| Red. to long. 77° 1.6′ W......= − 0.9 = | 1 32.1 p.m. |
| Hour angle of Polaris west of the meridian | =$4^h$ $47.8^m$ |
| Declination of Polaris | =88° 49′ 41″ N. |

| Mean time hour angle. | Azimuth of Polaris. Mean declination. +88° 49′ 45″ Latitude. 38° | 38° 54′ | 40° | Correction additive for declination +88° 49′ 40″ |
|---|---|---|---|---|
| $4^h$ $39.2^m$ | 84′.2 | 85′.3 | 86′.7 | 0′.1 |
| 47.8 | | 86 .3 | | 0 .1 |
| 49.2 | 85 .4 | 86 .5 | 87 .9 | 0 .1 |

Azimuth of Polaris=1° 26′.3+0′.1=1° 26′ 24″

### Final field notes.

March 21, 1910, at a transit point in Washington, D. C., in latitude 38° 53′ 40″ N., and longitude 77° 1′.6 W., I find by comparison with a Western Union telegraph clock that my watch is $1^m$ $22^s$ slow of 75th meridian standard time.

At the same station at $6^h$ $19.9^m$ p. m., l. m. t., I make an hour angle observation on Polaris, west of the meridian, two each with the telescope in direct and reversed positions, reading the horizontal deflection angle from a flag pole about 20 chs. S., in the direction S-W-N to Polaris.

Watch time of obsn.=$6^h$ $26^m$ $40^s$ p. m.

Mean horizontal angle from Polaris to flag =177° 34′ 15″ N-W-S

Azimuth of Polaris= 1 26 24 W.

True bearing of flag = N. 179° 00′ 39″ W.
= S. 0° 59′ 21″ W.

### POLARIS AT SUNSET OR SUNRISE.

99. Polaris is conveniently observed for azimuth by the hour angle method at sunset or sunrise without artificial illumination. The preparation for the observation consists in computing in advance the approximate settings in azimuth and altitude in order to find Polaris, and the plan contemplates an approximate reference meridian: With the time of sunset or sunrise assumed as the time of observation, the hour angle "$t$" and azimuth "$A$" are ascertained in order to find the position of Polaris in azimuth; the position in altitude is found by the following approximation, the positive sign being used for hour angles less than 6 hours and the negative sign for hour angles exceeding 6 hours:

$$v \neq \phi \pm 70' \cos t$$

Example of computation of the position of Polaris at sunset, May 6, 1911, at a station in latitude 47° 20′ N., and longitude 102° 40′ W.:

From the Ephemeris the declination of the sun is found to be 16° 18′ N., and by entering Table 17, of the Standard Field Tables, the apparent time of sunset is found to be $7^h\ 15^m$ p. m.

| | | | | | |
|---|---|---|---|---|---|
| Assumed time of obsn., May 6, 1911 | | | = | $7^h\ 15^m$ | p. m. |
| Gr. U. C. of Polaris, May 6 | = | $10^h\ 33.5^m$ a. m. | +12 | | |
| Red. to long. 102° 40′ W. | | −1.1 | = | 10 32.4 | a. m. |
| Assumed hour angle of Polaris west of the meridian | | | = | $8^h\ 42.6^m$ | |
| Hour angle, angular measure | | | = | 130° 39′ | |
| Azimuth of Polaris, W. | | | ≠ | 1° 17′ | |
| Latitude of station | = | 47° 20′ | | | |
| $70' \cos t = 70 \cos 130° 39'$ | = | 46(−) | | | |
| | | $v \neq 46° 34'$ | | | |

Example of computation of the position of Polaris at sunset, Nov. 6, 1911, at a station in latitude 47° 20′ N., and longitude 102° 40′ W.:

From the Ephemeris the declination of the sun is found to be 15° 44′ S., and by entering Table 17, of the Standard Field Tables, the apparent time of sunrise is found to be $7^h\ 12^m$ a. m., or of sunset $4^h\ 48^m$ p. m.

| | |
|---|---|
| Gr. U. C. of Polaris, Nov. 6, 1911 | $=10^h\ 28.2^m$ p. m. |
| Red. to long. 102° 40′ W. | $= \quad -1.1$ |
| L. M. T. of U. C. of Polaris | $=10^h\ 27.1^m$ p. m. |
| Assumed time of observation | $= 4 \quad 48$ p. m. |
| Assumed hour angle of Polaris east of the meridian | $= 5^h\ 39.1^m$ |
| Hour angle, angular measure | =84° 46′ |
| Azimuth of Polaris, E. | ≮ 1° 43′ |

| | |
|---|---|
| Latitude of station | =47° 20′ |
| 70′ cos $t$=70 cos 84° 46′ | = 06 (+) |
| | $v$≮47° 26′ |

Example of computation of the position of Polaris at sunrise, November 7, 1911, and same station as above:

| | |
|---|---|
| Assumed time of obsn., Nov. 7, 1911 | $=\begin{cases} 7^h\ 12^m \text{ a. m.} \\ +12 \end{cases}$ |
| L. M. T. of U. C. of Polaris, Nov. 6 | = 10 27.1 p. m. |
| Assumed hour angle of Polaris west of the meridian | $= 8^h\ 44.9^m$ |
| Hour angle, angular measure | = 131° 14′ |
| Azimuth of Polaris, W. | ≮ 1° 16′ |

| | |
|---|---|
| Latitude of station | =47°20′ |
| 70′ cos $t$=70 cos 131°14′= | 46(–) |
| | $v$≮46°34′ |

Thus at the above station in latitude 47° 20′ N., and longitude 102° 40′ W., to observe Polaris by the daylight method an approximate meridian should be established with the solar before sunset, then to *find* Polaris the following angles are set off:

| | Horizontal angle. | Vertical angle. |
|---|---|---|
| Sunset May 6, 1911 | 1°17′ W. | 46° 34′ |
| Sunset Nov. 6, 1911 | 1°43′ E. | 47° 26′ |
| Sunrise Nov. 7, 1911 | 1°16′ W. | 46° 34′ |

The above "settings" are merely approximations, but sufficiently close, however, to bring Polaris reasonably near the center of the field of the telescope where the star will be found in plain view; the telescope should be focused upon a distant object, otherwise, though Polaris may be practically at the center of the field, it might be out of focus and therefore not observable during daylight. When Polaris has been found the above settings have answered their purpose and the observation may proceed in accordance with either observing program "*a*" or "*b*" of the hour angle method, the final reductions to be based upon the precise details of the observation. During the reversals of the transit the settings should be made each time. The daylight hour angle method is particularly desirable because the observation, including all instrumental work, marking of points upon the ground, etc., is accomplished without artificial illumination, and sunset is usually a convenient time to devote to this field duty.

To recapitulate, the following general program will be found best adapted to the requirements of public-land surveying practice, and will be used most extensively:

Time: By meridian observation of the sun for apparent noon.

Latitude: By meridian altitude observation of the sun.

Azimuth, true meridian upon which to test the solar apparatus: By hour angle observation on Polaris at sunset.

Azimuth, on line: By the solar transit properly adjusted to the true meridian.

### ALTITUDE OBSERVATION OF THE SUN FOR AZIMUTH.

**100.** While the methods of observation upon Polaris for azimuth are unquestionably the most desirable in their relation to the theory and practice of public-land surveying, yet a very efficient alternative is found in direct altitude observations upon the sun for azimuth, with a number of equations at the disposal of the surveyor to suit his convenience. During the shorter days of the year and even quite often at any season the surveyor finds himself at a loss for time and suitable daylight hours in which to make the required tests of his solar attachment; conditions obtain making the required tests impossible if limited to a Polaris meridian in camp, without involving unreasonable delay. It is in such cases that a direct altitude observation upon the sun for azimuth, on the actual line of the survey, finds its most useful application. Presuming the surveyor at work with a standard instrument with solar attachment, the accuracy of its adjustments can, by this method, be readily tested at work on line at any suitable morning or afternoon hour, without appreciable loss of time. Under working conditions any line determined with the solar attachment may be used for reference purposes, while vertical and horizontal angles are recorded to the sun to obtain the necessary data for computing the true bearing of the established solar line. A series of three altitude observations upon the sun, each with the telescope in direct and reversed positions, are required to guard against error; these are readily made in 10 or 12 minutes, while the reductions may be made in the evening without loss of time from the line work.

Other difficulties in the nature of temporary disability of the solar attachment, and cloudy nights preventing Polaris observations, or other adverse conditions may sometimes obtain, during which periods, even for a few days, if the surveyor is familiar with the method of direct altitude observation upon the sun for meridian, he can thus establish his lines and possibly realize a saving of the entire time of his party until the trouble is removed. To the surveyors who have used this method little more needs to be said in its favor, but to those unfamiliar with it the suggestion is made to practice the observations and reductions until proficiency is attained, and in its application the reward will come many times during an average season's work.

Referring to the description of the standard instrument adopted by the General Land Office it will be noted that it is equipped with a full vertical circle, a colored glass shade in the dust shutter of the eye-piece, and a prismatic eye-piece; these are essential to rapid and accurate altitude observations upon the sun.

**101.** An altitude observation of the sun for azimuth consists in the simultaneous determination of the true vertical and horizontal angles to the sun's center, the horizontal angle being referred to a fixed point. With the true vertical angle to the sun's center, the declination of the sun, and the latitude of the station all known, one of the following equations is entered and a calculation made of the azimuth of the sun's center at the epoch of observation, as referred to the true meridian; the relation between the sun's calculated azimuth and the recorded angle to the sun's center gives the true bearing of the fixed reference point.

**102.** *Altitude observation of the sun for azimuth.*—Reverse the signs of "$\delta$" for south declinations:

$$\operatorname{Tan} \tfrac{1}{2} A=\sqrt{\frac{\cos \frac{1}{2}(\zeta+\phi+\delta) \sin \frac{1}{2}(\zeta+\phi-\delta)}{\cos \frac{1}{2}(\zeta-\phi-\delta) \sin \frac{1}{2}(\zeta-\phi+\delta)}}.$$

The spherical angles "$\zeta$", "$\phi$", and "$\delta$" appear in this equation combined as in the formula for the reduction of an altitude observation of the sun for apparent time, and when it is desired to reduce for both time and azimuth, the above equation for azimuth is to be preferred to any that follow.

**103.** *Altitude observation of the sun for azimuth.*—For south declinations the function "sin $\delta$" becomes negative by virtue of the sine of a negative angle being treated as negative in analytical reductions: If the algebraic sign of the result is positive the azimuth "$A$" is referred to the north point, but if negative, the azimuth "$A$" is referred to the south point:

$$\operatorname{Cos} A=\frac{\sin \delta}{\cos \phi \cos h}-\tan \phi \tan h$$

The above equation is very convenient in reducing for azimuth only.

**104.** *Altitude observation of the sun for azimuth.*—To many surveyors the following equation is familiarly expressed directly in terms of the spherical triangle "pole-zenith-sun:" Reverse the sign of "$\delta$" for south declinations:

Pole to zenith $=90^\circ-\phi=$ colat.;
Pole to sun $=90^\circ-\delta=$ codecl.;
Zenith to sun $=90^\circ-h=$ coalt.;
$S=\frac{1}{2}$ sum of the three sides:

$$\operatorname{Cos} \tfrac{1}{2} A=\sqrt{\frac{\sin S \sin (S-\text{codecl.})}{\sin \text{colat.} \sin \text{coalt.}}}$$

OBSERVING PROGRAM, MORNING.

**105.** Thoroughly level the transit.

With the telescope in direct position observe and record the horizontal deflection angle from a fixed reference point to the sun's right limb, and the vertical angle to the sun's upper limb; these observations must be simultaneous, at the epoch of which the sun will appear as indicated; note the watch time at the epoch of the observation: 

Reverse the transit.

Observe and record the horizontal deflection angle from the fixed reference point to the sun's left limb, and the vertical angle to the sun's lower limb; these observations must be simultaneous, at the epoch of which the sun will appear as indicated; note the watch time at the epoch of the observation: 

The mean observed vertical and horizontal angles, and the mean watch time are to be used in the reduction; this program constitutes one complete altitude observation, which is repeated until a series of three complete direct and reversed observations are made.

###### OBSERVING PROGRAM, AFTERNOON.

**106.** In the afternoon the program is modified only as to the order in which the sun's limbs are observed, which is as follows:

First observation, telescope direct, observe the sun's right and lower limbs: 

Second observation, telescope reversed, observe the sun's left and upper limbs: 

**107.** By the above observing programs the horizontal and vertical angles in the direct positions of the telescope will be found of about the same numerical values as in the reversed position of the telescope, by reason of the sun passing in a direction that will carry it across the field of the telescope during the time taken in the reversal and second setting. Differential refraction is therefore practically eliminated, and it is desirable that the corresponding angles in the direct and reversed positions of the telescope be about the same rather than as far apart as would result in any other observing program.

The most suitable hour for this observation is when the sun is moving rapidly in altitude as compared with a relatively small change in azimuth. When the sun has been brought into about the proper position in the field of the telescope the observer by lateral motion of the horizontal tangent screw on the plates keeps the vertical wire tangent to the sun's right or left limb while the upper or lower limb of the sun by the direction of its motion gradually approaches the horizontal wire; at the epoch of proper tangency of the two limbs to the two wires the observation is completed by calling "time" and stopping all motion until the angles are recorded. It is very helpful for an assistant to read the time and to enter all records.

**108.** Example of direct altitude observation of the sun for azimuth, sun north declination, and both north and south of an east and west line:

*Final field notes.*

Aug. 2, 1909, at the cor. of Tps. 31 and 32 S., Rs. 43 and 44 W., 6th Prin. Mer., Colo., in latitude 37° 17′.5 N., and longitude 102° 18′.6 W., at 7h 30m. a. m., app. t., I set off 37° 17′ 30″ N., on the lat. arc; 17° 52′ N., on the decl. arc; and determine a meridian with the solar, whence I turn 90° to the east and set a flag, about 20 chs. dist.; then to test this indication of the solar I make a series of three altitude observations of the sun for azimuth, each with the telescope in direct and reversed positions, observing opposite limbs of the sun, and reading the horizontal deflection angles from the flag to the sun:

| Observation. | Telescope. | Sun. | Watch time. | Vertical angle. | Horizontal angle from flag to sun. |
|---|---|---|---|---|---|
| 1st | Direct...... | ☉ | 7h 36m 54s | 30° 05′ | 0° 08′ 30″ to N. |
| " | Reversed... | ☉ | 7 38 15 | 29 48 | 0 33 00 " " |
| | Mean.. | ............. | ............. | 29° 56′ 30″ | 0° 20′ 45″ to N. |
| 2nd | Direct...... | ☉ | 7h 41m 20s | 30° 58′ 00″ | 0° 32′ 00″ to S. |
| " | Reversed... | ☉ | 7 43 00 | 30 46 30 | 0 12 30 " " |
| | Mean.. | ............. | ............. | 30° 52′ 15″ | 0° 22′ 15″ to S. |
| 3rd | Direct...... | ☉ | 7h 52m 00s | 33° 05′ 00″ | 2° 11′ 00″ to S. |
| " | Reversed... | ☉ | 7 53 48 | 32 53 30 | 1 50 00 " " |
| | Mean.. | ............. | 7h 52m 54s | 32° 59′ 15″ | 2° 00′ 30″ to S. |

| | |
|---|---|
| By 1st obsn. flag bears | N. 89° 58′ 57″ E. |
| By 2nd obsn. flag bears | N. 89 58 26 E. |
| By 3rd obsn. flag bears | N. 89 58 38 E. |
| Mean true bearing of flag | N. 89° 58′ 40″ E. |
| Indicated error of solar attachment | 1′ 20″ |

*Field record.*

The declination of the sun for the mean period of the three observations=17° 51′ 04″ N.

The following reductions are made to obtain the true vertical angles of the above observations:

| | 1st obsn. | 2nd obsn. | 3rd obsn. |
|---|---|---|---|
| $v$ | =29° 56′ 30″ | 30° 52′ 15″ | 32° 59′ 15″ |
| Refraction = | −1 40 | −1 36 | −1 28 |
| Parallax = | + 8 | + 8 | + 8 |
| $h$ | =29° 54′ 58″ | 30° 50′ 47″ | 32° 57′ 55″ |

The following examples of reduction are all by the equation:

$$\text{Cos } A=\frac{\sin \delta}{\cos \phi \cos h}-\tan \phi \tan h$$

| | | |
|---|---|---|
| log cos $\phi$=9.900674 | log sin $\delta$=9.486493(+) | log tan $\phi$=9.881708 |
| " cos $h$=9.937897 | | " tan $h$=9.759970 |
| 9.838571 | 9.838571 | log 9.641678 |
| | log 9.647922 | nat(−) .43821 |
| | nat (+) .44455 | |
| | (−) .43821 | |
| | cos $A$=(+) .00634 | |

$A$=True bearing of sun =N. 89° 38′ 12″ E.
Angle from sun to flag =(+) 0 20 45

True bearing of flag =N. 89° 58′ 57″ E.

| | | |
|---|---|---|
| log cos $\phi$=9.900674 | log sin $\delta$=9.486493(+) | log tan $\phi$=9.881708 |
| " cos $h$=9.933763 | | " tan $h$=9.776132 |
| 9.834437 | 9.834437 | log 9.657840 |
| | log 9.652056 | nat(−) .45482 |
| | nat(+) .44880 | (+) .44880 |
| | | cos $A$=(−) .00602 |

$A$=True bearing of sun =S. 89° 39′ 19″ E.
Angle from sun to flag =(+) 0 22 15

True bearing of flag =S. 90° 01′ 34″ E.
=N. 89° 58′ 26″ E.

| | | |
|---|---|---|
| log cos $\phi$=9.900674 | log sin $\delta$=9.486493(+) | log tan $\phi$=9.881708 |
| " cos $h$=9.923762 | | " tan $h$=9.811941 |
| 9.824436 | 9.824436 | log 9.693649 |
| | log 9.662057 | nat (−) .49391 |
| | nat (+) .45926 | (+) .45926 |
| | | cos $A$=(−) .03465 |

$A$=True bearing of sun =S. 88° 00′ 52″ E.
Angle from sun to flag =(+) 2 00 30

True bearing of flag =S. 90° 01′ 22″ E.
=N. 89° 58′ 38″ E.

The particular convenience of the above equation is noted in the fact that the functions "cos $\phi$", "tan $\phi$", and "sin $\delta$" are constant throughout the entire reduction, the function "$h$" being the only variable.

**109.** The third of the above series is selected for an example of reduction by the equation:

$$\text{Cos } \tfrac{1}{2} A = \sqrt{\frac{\sin S \sin (S - \text{codecl.})}{\sin \text{colat.} \sin \text{coalt.}}}$$

90°−$\phi$=90°−37° 17′ 30″ =52° 42′ 30″=colat.
90°−$\delta$=90°−17° 51′ 04″(+)=72 08 56 =codecl.
90°−$h$=90°−32° 57′ 55″ =57 02 05 =coalt.

2 $S$=181° 53′ 31″

$S$= 90° 56′ 45″
codecl.=90°−$\delta$= 72 08 56

$S$−codecl.= 18° 47′ 49″

| | | |
|---|---|---|
| log sin $S$ | = | 9.999941 |
| " sin ($S$−codecl.)= | | 9.508146 |
| " sin colat. | =9.900674 | 9.508087 |
| " sin coalt. | =9.923762 | |
| | 9.824436 | 9.824436 |
| " cos² ½$A$ | = | 9.683651 |
| " cos ½$A$ | = | 9.841825 |
| ½$A$ | = | 45° 59′ 35″ |

$A$=True bearing of sun =N. 91° 59′ 10″ E.
Angle from sun to flag =(−) 2 00 30

True bearing of flag =N. 89° 58′ 40″ E.

The above equation is as good as any for the reduction of one observation, but the reduction becomes laborious for a series of three observations.

**110.** The third of the above series is also selected for an example of reduction by the equation:

$$\operatorname{Tan} \tfrac{1}{2} A = \sqrt{\frac{\cos \frac{1}{2}(\zeta+\phi+\delta) \sin \frac{1}{2}(\zeta+\phi-\delta)}{\cos \frac{1}{2}(\zeta-\phi-\delta) \sin \frac{1}{2}(\zeta-\phi+\delta)}}$$

$$h = 32^\circ\ 57'\ 55''$$

| | |
|---|---|
| $\zeta = 57^\circ\ 02'\ 05''$ | $\zeta = 57^\circ\ 02'\ 05''$ |
| $\phi = 37\ \ 17\ \ 30$ | $\phi = 37\ \ 17\ \ 30$ |
| $\zeta+\phi = 94^\circ\ 19'\ 35''$ | $\zeta-\phi = 19^\circ\ 44'\ 35''$ |
| $\delta = 17\ \ 51\ \ 04\ (+)$ | $\delta = 17\ \ 51\ \ 04\ (+)$ |
| $\zeta+\phi+\delta = 112^\circ\ 10'\ 39''$ | $\zeta-\phi+\delta = 37^\circ\ 35'\ 39''$ |
| $\frac{1}{2}(\zeta+\phi+\delta) = 56^\circ\ 05'\ 20''$ | $\frac{1}{2}(\zeta-\phi+\delta) = 18^\circ\ 47'\ 50''$ |
| $\zeta+\phi = 94^\circ\ 19'\ 35''$ | $\zeta-\phi = 19^\circ\ 44'\ 35''$ |
| $\delta = 17\ \ 51\ \ 04\ (+)$ | $\delta = 17\ \ 51\ \ 04\ (+)$ |
| $\zeta+\phi-\delta = 76^\circ\ 28'\ 31''$ | $\zeta-\phi-\delta = 1^\circ\ 53'\ 31''$ |
| $\frac{1}{2}(\zeta+\phi-\delta) = 38^\circ\ 14'\ 15''$ | $\frac{1}{2}(\zeta-\phi-\delta) = 0^\circ\ 56'\ 45''$ |

| | | |
|---|---|---|
| $\log \cos \frac{1}{2}(\zeta+\phi+\delta) =$ | | 9.746561 |
| " $\sin \frac{1}{2}(\zeta+\phi-\delta) =$ | | 9.791636 |
| | | 9.538197 |
| " $\cos \frac{1}{2}(\zeta-\phi-\delta) =$ | 9.999941 | |
| " $\sin \frac{1}{2}(\zeta-\phi+\delta) =$ | 9.508152 | |
| | 9.508093 | 9.508093 |
| $\log \tan^2 \frac{1}{2}A =$ | | 0.030104 |
| " $\tan \frac{1}{2}A =$ | | 0.015052 |
| $\frac{1}{2}A =$ | | $45^\circ\ 59'\ 34''$ |

$A$ = True bearing of sun $=$ N. $91^\circ\ 59'\ 08''$ E.

Angle from sun to flag $= (-)2\ \ 00\ \ 30$

True bearing of flag $=$ N. $89^\circ\ 58'\ 38''$ E.

**111.** The above equation is as good as any for the reduction of one observation, but the reduction becomes laborious for a series of three observations. However, the advantage in using the above equation is found when it becomes desirable to reduce the observations for both time and azimuth.

Let it be required to reduce the third observation of the above series for time, making the reduction by the following equation:

$$\text{Tan}\ \tfrac{1}{2}\,t=\sqrt{\frac{\sin\frac{1}{2}(\zeta+\phi-\delta)\ \sin\frac{1}{2}(\zeta-\phi+\delta)}{\cos\frac{1}{2}(\zeta+\phi+\delta)\ \cos\frac{1}{2}(\zeta-\phi-\delta)}}$$

| | | |
|---|---|---|
| log $\sin\frac{1}{2}(\zeta+\phi-\delta)=$ | | 9.791636 |
| " $\sin\frac{1}{2}(\zeta-\phi+\delta)=$ | | 9.508152 |
| | | 9.299788 |
| " $\cos\frac{1}{2}(\zeta+\phi+\delta)=$ | 9.746561 | |
| " $\cos\frac{1}{2}(\zeta-\phi-\delta)=$ | 9.999941 | |
| | 9.746502 | 9.746502 |
| " $\tan^2\frac{1}{2}t$ | = | 9.553286 |
| " $\tan\ \frac{1}{2}t$ | = | 9.776643 |
| $\frac{1}{2}t$ | $=30^\circ\ 52'\ 34''$ | |
| $t$ | $=61^\circ\ 45'\ 08''$ | $=4^h\ 07^m\ 01^s$ |

| | |
|---|---|
| Apparent time of observation | $=7^h\ 52^m\ 59^s$ a. m. |
| Equation of time | $=\ +6\ \ 05$ |
| Local mean time of observation | $=7^h\ 59^m\ 04^s$ a. m. |
| Watch time of observation | $=7\ \ 52\ \ 54$ |
| Watch slow of l. m. t. | $=\ 6^m\ 10^s$ |

**112.** Example of direct altitude observation of the sun for azimuth, sun south declination:

*Final field notes.*

March 18, 1910, at a transit point in Washington, D. C., in latitude 38° 53′ 40″ N., and longitude 77° 01′.6 W., at $3^h\ 42^m$ p. m., app. t., I make a series of three altitude observations upon the sun for azimuth, each with the telescope in direct and reversed positions, observing opposite limbs of the sun, and reading the horizontal deflection angle from a flag pole about 20 chs. to the S., SW. to the sun:

| Obser-vation. | Telescope. | Sun. | Watch time. | Vertical angle. | Horizontal angle flag to sun. |
|---|---|---|---|---|---|
| 1st..... | Direct....... |  | $3^h\ 56^m\ 58^s$ | 25° 20′ | 65° 00′ to SW. |
| " ... | Reversed.... |  | 3 58 48 | 25 31 | 64 45 " " |
| | Mean ............ | | $3^h\ 57^m\ 53^s$ | 25° 25′ 30″ | 64° 52′ 30″ " " |
| 2nd.... | Direct....... |  | $4^h\ 01^m\ 48^s$ | 24° 28′ | 65° 56′ " " |
| " ... | Reversed.... |  | 4 03 10 | 24 44 | 65 36 " " |
| | Mean ............ | | ............ | 24° 36′ 00″ | 65° 46′ 00″ " " |
| 3rd.... | Direct....... |  | $4^h\ 05^m\ 58^s$ | 23° 44′ | 66° 44′ " " |
| " ... | Reversed.... |  | 4 07 30 | 23 57 | 66 26 " " |
| | Mean ............ | | ............ | 23° 50′ 30″ | 66° 35′ 00″ " " |

**By 1st obsn. flag bears S. 1° 00′ 02″ W.**
**" 2nd " " " S. 1 00 20 W.**
**" 3rd " " " S. 0 59 50 W.**

**Mean true bearing of flag=S. 1° 00′ 04″ W.**

*Field record.*

**The declination of the sun for the mean period of the three observations=1° 02′ 16″ S.**

**The following reductions are made to obtain the true vertical angles of the above observations:**

| | 1st obsn. | 2nd obsn. | 3rd obsn. |
|---|---|---|---|
| $v$= | 25° 25′ 30″ | 24° 36′ 00″ | 23° 50′ 30″ |
| Refraction= | −2 00 | −2 06 | −2 10 |
| Parallax= | + 08 | + 08 | + 08 |
| $h$= | 25° 23′ 38″ | 24° 34′ 02″ | 23° 48′ 28″ |

**113.** The first of the above series is selected for an example of reduction by the equation:

$$\operatorname{Tan} \tfrac{1}{2}A = \sqrt{\frac{\cos \frac{1}{2}(\zeta+\phi+\delta)\ \sin \frac{1}{2}(\zeta+\phi-\delta)}{\cos \frac{1}{2}(\zeta-\phi-\delta)\ \sin \frac{1}{2}(\zeta-\phi+\delta)}}$$

$$h = 25^\circ\ 23'\ 38''$$

$$\begin{aligned}
\zeta &= 64^\circ\ 36'\ 22'' \\
\phi &= 38\ \ 53\ \ 40 \\
\hline
\zeta+\phi &= 103^\circ\ 30'\ 02'' \\
\delta &= 1\ \ 02\ \ 16\ (-) \\
\hline
\zeta+\phi+\delta &= 102^\circ\ 27'\ 46'' \\
\tfrac{1}{2}(\zeta+\phi+\delta) &= 51^\circ\ 13'\ 53'' \\
\hline
\zeta+\phi &= 103^\circ\ 30'\ 02'' \\
\delta &= 1\ \ 02\ \ 16\ (-) \\
\hline
\zeta+\phi-\delta &= 104^\circ\ 32'\ 18'' \\
\tfrac{1}{2}(\zeta+\phi-\delta) &= 52^\circ\ 16'\ 09''
\end{aligned}$$

$$\begin{aligned}
\zeta &= 64^\circ\ 36'\ 22'' \\
\phi &= 38\ \ 53\ \ 40 \\
\hline
\zeta-\phi &= 25^\circ\ 42'\ 42'' \\
\delta &= 1\ \ 02\ \ 16\ (-) \\
\hline
\zeta-\phi+\delta &= 24^\circ\ 40'\ 26'' \\
\tfrac{1}{2}(\zeta-\phi+\delta) &= 12^\circ\ 20'\ 13'' \\
\hline
\zeta-\phi &= 25^\circ\ 42'\ 42'' \\
\delta &= 1\ \ 02\ \ 16\ (-) \\
\hline
\zeta-\phi-\delta &= 26^\circ\ 44'\ 58'' \\
\tfrac{1}{2}(\zeta-\phi-\delta) &= 13^\circ\ 22'\ 29''
\end{aligned}$$

$$\begin{aligned}
\log \cos \tfrac{1}{2}(\zeta+\phi+\delta) &= 9.796697 \\
\text{“}\ \sin \tfrac{1}{2}(\zeta+\phi-\delta) &= 9.898118 \\
\hline
 & \phantom{=}\ 9.694815 \\
\text{“}\ \cos \tfrac{1}{2}(\zeta-\phi-\delta) &= 9.988058 \\
\text{“}\ \sin \tfrac{1}{2}(\zeta-\phi+\delta) &= 9.329724 \\
\hline
 & \phantom{=}\ 9.317782 \qquad 9.317782 \\
\hline
\log \tan^2 \tfrac{1}{2}A &= 0.377033 \\
\text{“}\ \tan \tfrac{1}{2}A &= 0.188516 \\
\hline
\tfrac{1}{2}A &= 57^\circ\ 03'\ 44''
\end{aligned}$$

$A$ = True bearing of sun = N. 114° 07′ 28″ W.
Angle from sun to flag = (+)64 52 30

True bearing of flag = N. 178° 59′ 58″ W.
= S. 1° 00′ 02″ W.

**114.** Let it also be required to reduce the first observation of the above series for time, making the reduction by the following equation:

$$\text{Tan } \tfrac{1}{2} t = \sqrt{\frac{\sin \frac{1}{2}(\zeta+\phi-\delta) \sin \frac{1}{2}(\zeta-\phi+\delta)}{\cos \frac{1}{2}(\zeta+\phi+\delta) \cos \frac{1}{2}(\zeta-\phi-\delta)}}$$

| | | |
|---|---|---|
| log sin $\frac{1}{2}(\zeta+\phi-\delta)$= | | 9.898118 |
| " sin $\frac{1}{2}(\zeta-\phi+\delta)$= | | 9.329724 |
| | | 9.227842 |
| " cos $\frac{1}{2}(\zeta+\phi+\delta)$= | 9.796697 | |
| " cos $\frac{1}{2}(\zeta-\phi-\delta)$= | 9.988058 | |
| | 9.784755 | 9.784755 |
| log tan² $\frac{1}{2}$ $t$= | | 9.443087 |
| " tan $\frac{1}{2}$ $t$= | | 9.721544 |
| $\frac{1}{2}$ $t$=27° 46′ 29″ | | |

$t$=55° 32′ 58″=3ʰ 42ᵐ 12ˢ

| | | |
|---|---|---|
| Apparent time of obsn. | = | 3ʰ 42ᵐ 12ˢ p. m. |
| Equation of time | = | +8 17 |
| Local mean time of obsn. | = | 3ʰ 50ᵐ 29ˢ p. m. |
| Watch time of obsn. | = | 3 57 53 |
| Watch fast of l. m. t. | = | 7ᵐ 24ˢ |

75th meridian time of comparison with a Western

| | | |
|---|---|---|
| Union telegraph clock | = | 4ʰ 30ᵐ 00ˢ p. m. |
| Correction for longitude | = | −08 06 |
| L. M. T. of comparison | = | 4 21 54 |
| Watch time of comparison | = | 4 29 20 |
| Watch fast of l. m. t | = | 7ᵐ 26ˢ |

**115.** The second observation of the above series is selected for an example of reduction by the equation:

$$\text{Cos } A = \frac{\sin \delta}{\cos \phi \cos h} - \tan \phi \tan h$$

| | | | | | |
|---|---|---|---|---|---|
| log cos $\phi$=9.891149 | log sin $\delta$= | 8.257958 (−) | log tan $\phi$= | 9.906733 |
| " cos $h$=9.958790 | | | " tan $h$= | 9.660053 |
| 9.849939 | | 9.849939 | log | 9.566786 |
| | log | 8.408019 | nat (−) | .36880 |
| | nat (−) | .02559 | (−) | .02559 |
| | | | cos $A$=(−) | .39439 |

| | | |
|---|---|---|
| $A$=True bearing of sun | = | S. 66° 46′ 20″ W. |
| Angle from sun to flag | = | (−)65 46 00 |
| True bearing of flag | = | S. 1° 00′ 20″ W. |

**116.** The third observation of the above series is selected for an example of reduction by the equation;

$$\text{Cos}\ \tfrac{1}{2}\ A=\sqrt{\frac{\sin S \sin (S-\text{codecl.})}{\sin \text{colat.} \sin \text{coalt.}}}$$

$$
\begin{array}{lll}
90^\circ-\phi=90^\circ-38^\circ\ 53'\ 40'' & = & 51^\circ\ 06'\ 20''=\text{colat.} \\
90^\circ-\delta=90^\circ-\ 1\ \ 02\ \ 16\ (-) & = & 91\ \ 02\ \ 16\ =\text{codecl.} \\
90^\circ-h=90^\circ-23\ \ 48\ \ 28 & = & 66\ \ 11\ \ 32\ =\text{coalt.} \\
\hline
 & 2\ S= & 208^\circ\ 20'\ 08'' \\
\hline
 & S= & 104^\circ\ 10'\ 04'' \\
\text{codecl.}\ =90^\circ-\delta & = & 91\ \ 02\ \ 16 \\
\hline
S-\text{codecl.} & = & 13^\circ\ 07'\ 48''
\end{array}
$$

$$
\begin{array}{llll}
\log \sin S & = & & 9.986585 \\
\text{"}\ \sin (S-\text{codecl.}) & = & & 9.356334 \\
\hline
 & & & 9.342919 \\
\text{"}\ \sin \text{colat.} & = & 9.891149 & \\
\text{"}\ \sin \text{coalt.} & = & 9.961376 & \\
\hline
 & & 9.852525 & 9.852525 \\
\hline
\log \cos^2 \tfrac{1}{2} A & = & & 9.490394 \\
\text{"}\ \cos \tfrac{1}{2} A & = & & 9.745197 \\
\tfrac{1}{2} A & = & & 56^\circ\ 12'\ 35'' \\
A=\text{True bearing of sun} & = & \text{N.}\ 112^\circ\ 25'\ 10''\ \text{W.} & \\
\text{Angle from sun to flag} & = & (+)\ 66\ \ 35\ \ 00 & \\
\hline
\text{True bearing of flag} & = & \text{N.}\ 179^\circ\ 00'\ 10''\ \text{W.} & \\
 & = & \text{S.}\ \ 0^\circ\ 59'\ 50''\ \text{W.} &
\end{array}
$$

EQUAL ALTITUDE OBSERVATIONS OF THE SUN FOR MERIDIAN.

**117.** The true meridian may be established by the method of equal altitude observations of the sun. The observation is not well adapted to line work, but it possesses a certain usefulness in camp, in that the surveyor may thus determine the true meridian by the sun with mere approximations as to time and latitude.

The fixation of the true meridian by this method depends upon the theory that the sun's center at equal altitudes occupies symmetrical positions in azimuth east and west of the meridian in the morning and in the afternoon except for the correction necessary to be applied due to the change in the sun's declination in the interval between the a. m. and p. m. observations:

"$dA_\delta$": Correction in azimuth in minutes of angular measure to be applied to the mean position in azimuth to obtain the true south point; the correction is to be applied to the east with a northerly hourly change in declination, or to the west with a southerly hourly change.

"$d\delta$": Change in declination of the sun from the a. m. to the p. m. observation, expressed in minutes of angular measure.

"$(t_1+t_2)$": The sum of the hour angles from apparent noon, or the total watch time from the a. m. to the p. m. observation, expressed in angular measure.

$$dA_\delta = \frac{\frac{1}{2}d\delta}{\cos\phi \sin\frac{1}{2}(t_1+t_2)}$$

The symmetry of the equal altitude observation is maintained by observing opposite limbs in azimuth in the a. m. and p. m. observations, in connection with the same limb in vertical angle in both observations.

With "$\frac{1}{2}d\delta$" and "$\frac{1}{2}(t_1+t_2)$" calculated, the computation can be concluded by applying to "$\frac{1}{2}d\delta$" the declination coefficient obtained by entering Table 22 of the Standard Field Tables, which gives coefficients for computing errors in azimuth due to small errors in declination, arguments: "$\phi$" and "$\frac{1}{2}(t_1+t_2)$."

**118.** An equal altitude observation of the sun for azimuth consists in reading the horizontal deflection angles from a fixed reference point to opposite right or left limbs of the sun in a. m. and p. m. observations simultaneously with the same upper or lower limb at the epoch of equal vertical angle in both observations, from the record of which a calculation is made of the bearing of the reference point as referred to the true meridian. To guard against error the surveyor is required to make a series of three equal altitude observations, taking the resulting mean. The most suitable a. m. and p. m. hours for this observation obtain when the sun is moving rapidly in altitude as compared with a relatively small change in azimuth.

EQUAL ALTITUDE OBSERVATIONS OF THE SUN, OBSERVING PROGRAM.

**119.** Select the observing station, or transit point, and a reference point preferably to the south, and not nearer than 5 or 10 chains distant.

Thoroughly level the transit for the a. m. observation.

Observe and record the horizontal deflection angle from the fixed reference point to the sun's right limb, and the vertical angle to the sun's lower limb; these observations must be simultaneous, at the epoch of which the sun will appear as indicated; note the watch time at the epoch of the observation: 

Thoroughly level the transit for the p. m. observation.

With the same vertical angle set off for the p. m. observation follow the sun's left limb until the sun's lower limb becomes tangent, as indicated, recording the watch time and horizontal deflection angle from the reference point: 

The above program constitutes one observation. A series of three observations are taken by three successive a. m. settings at intervals of about four or five minutes of time. In the p. m. the settings are of course made in the inverse order.

Consider each equal altitude observation separately and subtract the lesser horizontal angle from the greater and divide by two.

The mean of the three half-differences is then taken to determine the horizontal angle from the reference point to an *uncorrected* south point, this angle to be applied in a direction to equalize the south point between the two observed positions of the sun.

Compute the differential azimuth correction due to the change in the sun's declination from the mean period of the a. m. to the mean period of the p. m. observations, and apply this angle to the mean of the half-differences as stated above; the differential azimuth correction is to be applied to the east when the hourly change in the sun's declination is northerly or to the west when the hourly change in the sun's declination is southerly; the computed resultant angle indicates the bearing of the reference point referred to the true meridian.

The correct apparent times of the observations do not need to be known, as the function "$\frac{1}{2}(t_1+t_2)$" equals one-half the time in hours and minutes, by the surveyor's watch, from the a. m. to the p. m. observation.

The equal altitude observation may be modified by taking a p. m. observation one day followed by an a. m. observation the next, in which case the functions "$\frac{1}{2}d\,\delta$" and "$\frac{1}{2}(t_1+t_2)$" are to be computed for the period from the p. m. to the a. m. observation; and the differential azimuth correction, "$d\ A_s$", is then applied in the *opposite* direction.

**120.** Example of equal altitude observation of the sun for azimuth:

*Final field notes.*

**May 3, 1913, at a transit point in Washington, D. C., in latitude 38° 53′ 40″ N., and longitude 77° 1′.6 W., at 9ʰ27ᵐ a. m. and 2ʰ33ᵐ p. m., app. t., I make a series of three equal altitude observations upon the sun for azimuth, reading the horizontal deflection angle from a flag-pole about 20 chs. to the S., SE. in the a. m. to the sun's right limb, and SW. in the p. m. to the sun's left limb; equal vertical angles being taken to the sun's lower limb.**

| Observation. | Sun. | Watch time. | Vertical angle. | Horizontal angle flag to sun. | |
|---|---|---|---|---|---|
| 1st a. m. | q- | $9^h29^m25^s$ | 48°28′00″ | 67°20′00″ | to SE. |
| 3d p. m. | -p | 2 41 40 | | 65 28 30 | to SW. |
| | | | | 1°51′30″ (Diff.) | |
| 2d a. m. | q- | $9^h32^m50^s$ | 49°05′00″ | 66°29′30″ | to SE. |
| 2d p. m. | -p | 2 38 15 | | 64 38 00 | to SW. |
| Sum of hour angles | | $5^h05^m25^s$ | | | |
| Mean hour angle | | $2^h32^m42^s$ | | 1°51′30″ (Diff.) | |
| 3d a. m. | q- | $9^h36^m30^s$ | 49°43′00″ | 65°34′30″ | to SE. |
| 1st p. m. | -p | 2 34 45 | | 63 45 30 | to SW. |
| | | | | 1°49′00″ (Diff.) | |

One-half differences, or bearing angles from uncorrected south point to flag:

By 1st obsn. = S. 0° 55′ 45″ W.
" 2d " = S. 0 55 45 W.
" 3d " = S. 0 54 30 W.

Mean = S. 0° 55′ 20″ W.
Differential azimuth correction = (+) 3′ 53″

Mean true bearing of flag = S. 0° 59′ 13″ W.

*Field record.*

The hourly change in the sun's declination=44″.3 N.

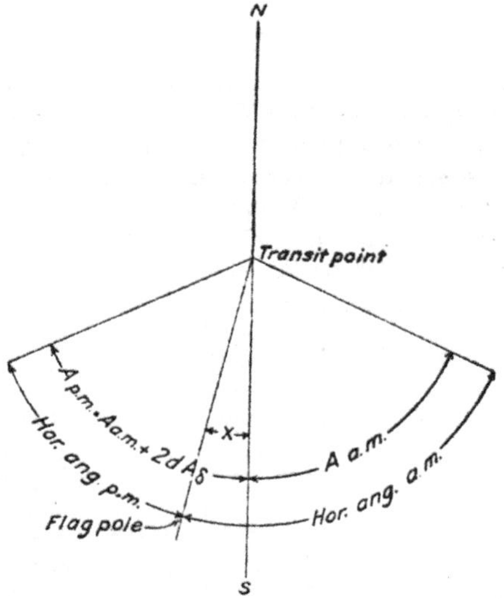


$$\text{Hor. ang. a. m.} = A \text{ a. m.} + x = A + x$$

$$\text{Hor. ang. p. m.} = A \text{ p. m.} - x = A + 2dA_\delta - x$$

$$\text{Hor. ang. a. m.} - \text{Hor. ang. p. m.} = 2x - 2dA_\delta$$

$$x = \frac{\text{Hor. ang. a. m.} - \text{hor. ang. p. m.}}{2} + dA_\delta$$

The following computation is made to obtain the differential azimuth correction for the above series:

$\frac{1}{2}\, d\delta = \frac{1}{2} \times 5.08 \times 44.3 = 112''$; $\log \frac{1}{2} d\delta = 2.049218$

$\phi = 38°\ 53'\ 40''$ N.; $\log \cos \phi \quad = 9.891149$

$\frac{1}{2}\,(t_1 + t_2) = 2^h\ 32^m\ 42^s$

$= 38°\ 10'\ 30''$

$\log \sin \frac{1}{2}(t_1 + t_2) \quad = 9.791034$

9.682183     9.682183

$\log d\, A_\delta = 2.367035$

$d\, A_\delta$ = Differential azimuth correction = 233″

= 3′ 53″

**121.** The following reduction to obtain the value of the differential azimuth correction for the above series is made with the use of Table 22 of the Standard Field Tables:

| Latitude. | $\frac{1}{2}(t_1+t_2)$, or hours from noon. | | |
|---|---|---|---|
| | $2^h$ | $2^h\ 33^m$ | $3^h$ |
| 35° 00′ | 2.44 | 2.05 | 1.73 |
| 38 54 | | 2.16 | |
| 40 00 | 2.61 | 2.19 | 1.85 |

Declination coefficient $=2.16$
$d\ A_\delta=2.16\times\frac{1}{2}\ d\delta=2.16\times112''$ $=242''$
$d\ A_\delta$=differential azimuth correction$=4'\ 02''$

The small difference (09″) in the computation of "$d\ A_\delta$" in the two processes of reduction is due to the error in adopting a coefficient obtained by linear interpolation in Table 22 of the Standard Field Tables, the tabular interval of which is large. Ordinarily the equal altitude method would be used when the latitude of the station is uncertain, and the slight error in using the declination coefficient taken by linear interpolation from Table 22 is small enough to be negligible.

**122.** The second a. m. and p. m. observations of the above series are selected for an example of reduction to the sun's center and direct computation of the sun's azimuth, and true bearing of the flag, by the equation:

$$\text{Cos } A=\frac{\sin \delta}{\cos \phi \cos h}-\tan \phi \tan h$$

| | |
|---|---|
| Vertical angle to sun's lower limb | =49° 05′ 00″ |
| Reduction to sun's center | =+ 15′ 54″ |
| Refraction | =− 49″ |
| Parallax | =+ 06″ |
| Sun's center, $h$ | =49° 20′ 11″ |

Declination of the sun at Greenwich apparent noon $=15° 34' 37''$ N.
Diff. in time to a. m. obsn.:

For longitude $= 5^h 08^m$
For time, a. m. $= -2\ 33$

$2.58^h = 2^h 35^m$

Diff. in declination to app. t. of a. m. obsn.:

$2.58 \times 44''.3 = 114'' = 1' 54''$ N.

Sun's decl. a. m. obsn. $= 15° 36' 31''$ N.
Diff. to p.m. obsn., already computed $(2 \times 112'' = 224'') = 3' 44''$ N.

Sun's decl. p. m. obsn. $= 15° 40' 15''$ N.

| | | a. m. obsn. | | p. m. obsn. |
|---|---|---|---|---|
| log cos $\phi = 9.891149$ | log sin $\delta = 9.429856(+)$ | | log sin $\delta = 9.431541(+)$ | |
| " cos $h = 9.813992$ | | | | |
| 9.705141 | | 9.705141 | | 9.705141 |
| " tan $\phi$ = 9.906733 | log | 9.724715 | log | 9.726400 |
| " tan $h$ = 0.065991 | nat (+) | .53054 | nat (+) | .53260 |
| log 9.972724 | | | | |
| nat (−) .93913 | (−) | .93913 | (−) | .93913 |
| cos $A$ = | (−) | .40859 | (−) | .40653 |
| $A$ = true bearing of sun | = | S. 65° 53′ 02″ E. | | S. 66° 00′ 47″ W. |

| | a. m. | p. m. |
|---|---|---|
| Horizontal angle from flag to sun's right and left limbs = | 66° 29′ 30″ to SE. | 64° 38′ 00″ to SW. |
| Reduction to sun's center $= \frac{15.9'}{\cos 49° 20'}$ = | (+) 24′ 24″ | (+) 24′ 24″ |
| Hor. ang. to sun's center = | 66° 53′ 54″ to SE. | 65° 02′ 24″ to SW. |
| Sun's azimuth as computed above | =S. 65° 53′ 02″ E. | S. 66° 00′ 47″ W. |
| True bearing of flag | =S. 1° 00′ 52″ W. | S. 0° 58′ 23″ W. |
| Mean true bearing of flag | =S. 0° 59′ 37″ W. | |

The discrepancy between the a. m. and p. m. results suggests a systematic instrumental error ordinarily eliminated by taking direct and reversed observations, which in this instance is of opposite effect in a. m. and p. m. hours and apparently eliminated in the mean result.

**123.** One additional fact should be noted relative to the several reductions of the above equal altitude observations:

$$\text{By above direct computation, } A \text{ p. m.} = 66° \ 00' \ 47''$$
$$A \text{ a. m.} = 65° \ 53' \ 02''$$
$$\text{Difference} = 2dA_\delta = 7' \ 45''$$
$$dA_\delta = 3' \ 53''$$

This value for $dA\delta$ (3′ 53″) agrees with same function as first computed.

**124.** Upon concluding the subject of azimuth determinations it will be of interest to note that the weighted mean of a large number of observations gives a value of S. 0° 59′ 25″ W. for the azimuth of the line from the Washington, D. C., transit point to the flag pole heretofore described. A comparison of the methods and results of the various observations as given on the preceding pages suggests that the surveyor should seldom be without means by which accurately to determine time, latitude and azimuth at any place in the field, however remote, and should doubt arise as to his results a "check" by independent method is nearly always available and a certain guide as to the accuracy of the determinations. It might be added that a careful surveyor will not fail to surround his methods with adequate verification to insure the accuracy required in the execution of the public-land surveys.

## THE TRUE PARALLEL OF LATITUDE.

**125.** The base lines and standard parallels of the rectangular system are established on the true parallel of latitude; the random latitudinal township boundary lines are also projected on the same curve; this curve is defined by a plane at right angles to the earth's polar axis cutting the earth's surface on a small circle. At every point on the true parallel the curve bears due east and west, the direction of the line being at right angles to the meridian at every point along the line. Two points at a distance of 20 chains apart on the same parallel of latitude may be said to define the direction of the curve at either point, without appreciable error, but the projection of a line so defined in either direction, easterly or westerly, would describe a great circle of the earth gradually departing southerly from the true parallel. The great circle tangent to the parallel at any origin or reference point along the parallel is known as the "tangent to the parallel," and it is coincident with the true latitude curve only at the point of origin. The rate of the change of the azimuth of the tangent is a function of the latitude on the earth's surface. The azimuth of the tangent varies directly as the distance from the origin, and the offset distance from the tangent to the parallel varies as the square of the distance from the point of tangency. A great circle connecting two distant points on the same latitude curve has the same angle with the meridian at both points and the azimuth of such a line at the two points of intersection is a function of one-half the distance between the points.

There are three general methods of establishing a true parallel of latitude which may be employed independently to arrive at the same result: (1) The solar method; (2) the tangent method; and, (3) the secant method.

SOLAR METHOD.

**126.** The solar instruments are capable of following the true parallel of latitude without substantial offsets. If such an instrument, in good adjustment, is employed, the true meridian may be determined by observation with the solar at each transit point. A turn of 90° in either direction then defines the true parallel, and if sights are taken not longer than from 20 to 40 chains distant, the line so established does not appreciably differ from the theoretical parallel of latitude. The locus of the resulting line is a succession of points each one at right angles to the true meridian at the previous station. However, during a period each day the solar is not available, and during this time, also whenever the sun may be obscured by clouds, or on account of a disturbance of the adjustments of the solar attachment, and whenever an instrument without solar attachment is employed, reference must be made to a transit line from which to establish the true latitude curve by one of the following methods.

TANGENT METHOD.

**127.** The tangent method of determination of the true latitude curve consists in establishing the true meridian at the point of beginning, from which a horizontal deflection angle of 90° is turned to the east or west, as may be required, and the projection of the line thus determined is called the tangent. The tangent is projected 6

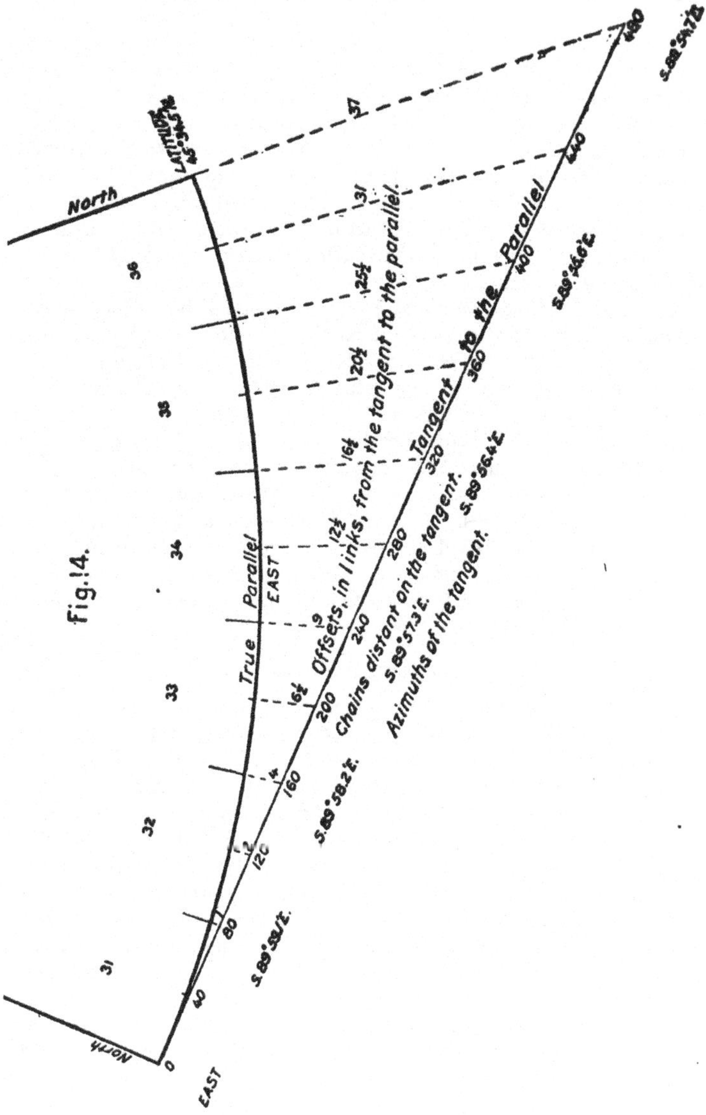
North
LATITUDE
True Parallel
EAST
Offsets, in links, from the tangent to the parallel.
Tangent to the Parallel
Chains distant on the tangent.
Azimuths of the tangent.
S.89°58.2'E.
S.89°57.3'E.
S.89°56.4'E.
S.89°55.6'E.
31
32
33
34
35
36
0
40
80
120
160
200
240
280
320
360
400
440
480
4
6¼
9
12¼
16¼
20¼
25¼
31
37
North
EAST

Fig. 14.

miles in a straight line, and as the measurements are completed for each corner point, proper offsets are measured north from the tangent to the parallel, upon which line the corners are established.

In Table 12, Standard Field Tables, are given the bearing angles or azimuths of the tangent to the parallel, referred to the true S. point, tabulated for any degree of latitude from 25° to 70° N., for the end of each mile from 1 to 6 miles. At the point of beginning the tangent bears east or west, but as the projection of the tangent is continued the deviation to the south increases in accordance with rules already stated.

In Table 13, Standard Field Tables, are shown the various offsets from the tangent north to the parallel, tabulated for any degree of latitude from 25° to 70° N., for each half mile from ½ to 6 miles.

The accompanying diagram illustrates the establishment of a standard parallel in latitude 45° 34′.5 N., by the tangent method. (See Fig. 14.) The form of record is shown in the specimen field notes.

Objection to the use of the tangent method in a timbered country is found owing to the requirement that all blazing is to be made on the true surveyed lines. Also, all measurements to items of topography entered in the field notes are to be referred to the true established lines. These objections to the tangent method, on account of the increasing distance from the tangent to the parallel, are largely removed in the secant method.

#### SECANT METHOD.

**128.** The designated secant is a great circle which cuts any true parallel of latitude at the first and fifth mile corners, and is tangent to an imaginary latitude curve at the third mile point. From the point of beginning to the third mile corner the secant has a northeasterly or northwesterly bearing; at the third mile corner the secant bears east or west; and from the third to the sixth mile corners the secant has a southeasterly or southwesterly bearing, respectively, depending upon the direction of projection, east or west. From the point of beginning to the first mile corner and from the fifth to the sixth mile corners the secant lies south of the true parallel, and from the first to the fifth mile corners the secant lies north of the true parallel. It will thus be seen that the secant method is a mere modification of the tangent method, so arranged that the minimum offsets can be made from the projected transit line to the established true parallel of latitude.

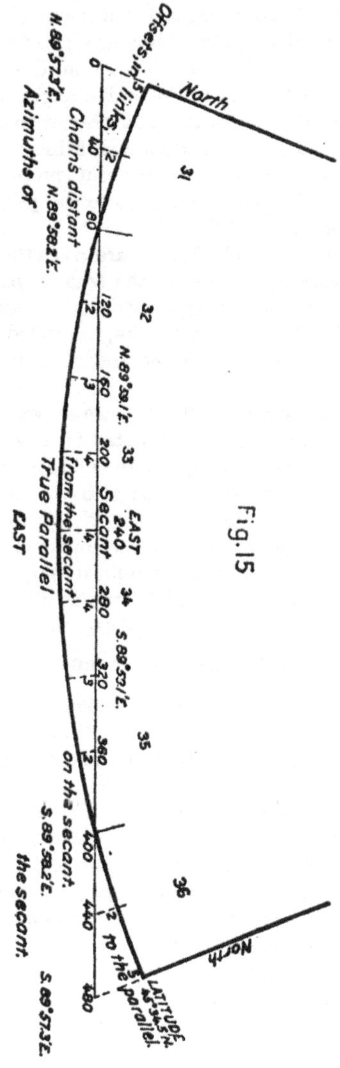
North
North
31
32
33
34
35
36
Offsets, in links
Chains distant
from the secant
on the secant.
to the parallel.
Secant
True Parallel
EAST
EAST
Azimuths of
the secant.
N. 89°57.3'E.
N. 89°58.2'E.
N. 89°59.1'E.
S. 89°59.1'E.
S. 89°58.2'E.
S. 89°57.3'E.
LATITUDE 45°34.5'N.

Fig. 15

The secant method of determination of the true latitude curve consists in establishing the true meridian at a point south of the beginning corner a measured distance taken from the table, from which meridian the proper horizontal deflection angle, as taken from the table, is turned to the northeast or northwest to define the secant. The secant is projected 6 miles in a straight line, and as the measurements are completed for each corner point, proper offsets are measured, north or south, from the secant to the parallel, upon which parallel the corners are established.

In Table 14, Standard Field Tables, are given the bearing angles or azimuths of the secant, referred to the true N. point for the first 3 miles, and the same symmetrical bearing angles or azimuths referred to the true S. point for the last 3 miles, tabulated for any degree of latitude from 25° to 70° N., for the end of each mile from 0 to 6 miles.

In Table 15, Standard Field Tables, are shown the various offsets from the secant to the parallel, tabulated for any degree of latitude from 25° to 70° N., for each half mile from 0 to 6 miles.

The accompanying diagram illustrates the establishment of a standard parallel in latitude 45° 34′.5 N. by the secant method. (See Fig. 15.) The form of record is shown in the specimen field notes.

The secant method is recommended for its simplicity of execution and proximity to the true latitude curve, as all measurements and cutting by this method are substantially on the true parallel.

### CONVERGENCY OF MERIDIANS.

**129.** The linear amount of the convergency of two meridians is a function of their distance apart, of the length of the meridian between two reference parallels, of the latitude, and of the spheroidal form of the earth's surface.

The following equation is convenient for the analytical computation of the linear amount of the convergency on the parallel, of two meridians any distance apart, and any length. The correction for convergency in any closed figure is proportional to the area, and may be computed from an equivalent rectangular area:

"$m_\lambda$": Measurement along the parallel.
"$m_\phi$": Measurement along the meridian.
"$a$": Equatorial radius of the earth=3963.3 miles.
"$e$": Factor of eccentricity, log $e$=8.915 2515.

"$dm_\lambda$": Linear amount of the convergency on the parallel, of two meridians distance apart "$m_\lambda$," and length "$m_\phi$" along the meridian: "$dm_\lambda$", "$m_\lambda$", "$m_\phi$" and "$a$" to be expressed in the same linear unit:

$$dm_\lambda = \frac{m_\lambda \, m_\phi}{a} \tan \phi \sqrt{1 - e^2 \sin^2 \phi}$$

Example of computation of the convergency of two meridians 24 miles long and 24 miles apart in a mean latitude of 43° 20′:

| | | |
|---|---|---|
| nat 1 | = | 1.0000000 |
| log $e$ | =8.915 2515 | |
| " " | =8.915 2515 | |
| " sin 43° 20′ | =9.836 477 | |
| " " " " | =9.836 477 | |
| " $e^2 \sin^2 \phi$ | =7.503 457 | |
| nat " " " | = | 0.0031875 |
| " $(1 - e^2 \sin^2 \phi)$ | = | 0.9968125 |
| log " " " | 9.998 614 | |
| " $\sqrt{1 - e^2 \sin^2 \phi}$ | =9.999 307 | |
| " tan 43° 20′ | =9.974 720 | |
| " 24 | =1.380 211 | |
| " " | =1.380 211 | |
| " 80* | =1.903 090 | |
| " product | =4.637 539 | |
| " 3963.3 | =3.598 057 | |
| " $dm_\lambda$ | =1.039 482 | |
| nat " | 10.9517 chs. | |

The convergency, measured on the parallel, of two meridians 24 miles apart and 24 miles long, in a mean latitude of 43° 20′, is therefore found to be 10.95 chains. The convergency of the east and west boundaries of a regular township in the same latitude would be equal to one-sixteenth of the convergency of the east and west boundaries of the quadrangle as computed above, or 68.44 links, which agrees with the value taken from Table 11 of the Standard Field Tables.

*This factor is introduced here for the purpose of conversion from the unit expressed in miles to the unit expressed in chains.

**130.** In Table 11, Standard Field Tables, are tabulated the linear amounts of the convergency of meridians, 6 miles long and 6 miles apart, for each degree of latitude from 25° to 70° N., together with the angle of convergency of the same meridians. These amounts of linear convergency are at once the proper corrections to apply to the north boundary of a regular township in the computation of the closing error around a township, or other computation by which a theoretical length of a north or south boundary of a township is compared with the length of the opposite boundary; the tabulated linear amounts of convergency are equal to double the amounts of the offsets from a tangent to the parallel at 6 miles for the same latitudes. Simple interpolation may be made for any intermediate latitude, and the amount of the convergency for a fractional township or other figure may be taken in proportion to the tabulated convergency as the fractional area is to 36 square miles.

The tabulated angle of convergency represents at once the deviation in azimuth of the tangent from the parallel at 6 miles; and $\frac{1}{6}$, $\frac{1}{3}$, $\frac{1}{2}$, $\frac{2}{3}$, and $\frac{5}{6}$ of the tabulated angles of convergency represent at once the amounts of the correction in the bearing of meridional section lines to compensate for convergency within a township.

In the same table are given the differences of longitude for 6 miles in both angular and time measure, also the differences of latitude, for 1 or 6 miles, in angular measure, in the various tabulated latitudes.

**131.** In the plan of subdivision of townships the meridional section lines are established parallel to the east boundary or other governing line; this necessitates a slight correction on account of the angular convergency of meridians. Meridional section lines west of the governing line are deflected to the left of the bearing of the governing line the amount shown in the second part of Table 2, Standard Field Tables, which is entered under two arguments: (1) Latitude, and (2) distance from the governing line. Meridional section lines east of a governing boundary are given the same amount of correction for bearing, but the deflection is made to the right.

## LENGTHS OF ARCS OF THE EARTH'S SURFACE.

**132.** All computations involving a difference of latitude for a given measurement along a meridian or the converse calculation, or other computations involving a difference of longitude for a given measurement along a parallel or a similar converse calculation, are readily accomplished by the use of the values given in Table 16, Standard Field Tables; this table gives the lengths in miles and decimal part of a mile of one degree of longitude measured on the parallel, and the lengths in miles of one degree of latitude measured on the meridian, for any latitude from 25° to 70° N.

The above tabulated values may be reduced to miles and chains, or to chains or feet, as convenient. In taking out lengths of degrees of longitude measured on the parallel an exact linear interpolation may be made, and in taking out lengths of degrees of latitude measured on the meridian the value should be taken out for the mean position in latitude of that portion of the meridian whose length it is desired to compute.

**133.** The first part of Table 2, Standard Field Tables, has been arranged for the reference of the latitude of any point within a township to the south boundary, the only argument being the miles and chains distant from the south boundary. Thus with the use of this table all observations for latitude within a township may be reduced to the south boundary; and conversely, given the latitude of the south boundary of a township, the latitude of any station within the township may readily be obtained by applying the difference given in the table for the known distance north.

# CHAPTER III.

## SYSTEM OF RECTANGULAR SURVEYS.

### GENERAL SCHEME.

**134.** In the preceding chapters there has been outlined the system of nomenclature and procedure relating, in general terms, to the survey of the public domain. It is confidently assumed that the United States surveyor has become impressed with the purpose of his task and the stability and dignity which should be attached to a work so great and important, commensurate with its broad foundation in law and science.

For the purpose of disposal of the public domain the law provides, in general terms, for its description, subdivision and identification in conformity with the following general scheme:

1st. The township, 6 miles square, containing 36 sections, each 1 mile square.

2d. The numbering of the townships meridionally into a range and latitudinally into a tier, from which the necessity at once appears for the selection of independent initial points, each to serve as an origin for the extension of surveys synchronously needed in somewhat widely separated localities, to provide for which, principal or governing meridians and base lines have been established, to which might be related the surveys executed in each of such localities.

3d. The establishment of guide meridians and correction lines or standard parallels at intervals sufficiently near each other to maintain a practical workable adherence to the legal definition of the primary unit, the township 6 miles square, and at the same time to reduce to a minimum the number of corners required.

4th. The placing of fractional sections on the north and west boundaries of the township.

5th. The subdivision of the townships into 36 sections by running parallel lines through the township from south to north and from east to west at distances of 1 mile.

6th. The inflexible declaration of the integrity of the corners marked in the public surveys as the proper legal corners of the sections or of the subdivisions of the sections which they were intended to designate, together with the equally important provisions (*a*) that the boundary lines actually run and marked shall be and remain the proper boundary lines of the sections or subdivisions for which they were intended; (*b*) that the length of such lines as returned by the surveyors shall be held as the true length thereof; and (*c*) that the sections shall be subdivided by running straight lines from the established quarter-section corners to the opposite established quarter-section corners.

**135.** The townships will be numbered to the north or south commencing with number 1 at the base line, and with range numbers to the east or west beginning with number 1 at the principal meridian.

The 36 sections into which a township is subdivided are numbered commencing with number 1 in the northeast section of the township, proceeding thence west to section 6, thence south to section 7, thence east to section 12, and so on, alternately, to number 36 in the southeast section. In the case of fractional townships, the sections will bear the same numbers they would have had if the townships were full, that is to say the section numbers should be employed which are the proper section numbers relating to the sides which are the governing boundaries, leaving any deficiency to fall on the opposite sides.

**136.** The specimen field notes will serve to illustrate the method of running lines to form quadrangles 24 miles square; the method of running the exterior lines of townships; and the method of subdividing regular townships. The methods here presented are designed to insure a full compliance with every practicable requirement, meaning and intent of the surveying laws.

**137.** By the terms of the original law and by general practice section lines are surveyed from south to north and from east to west, in order uniformly to place excess or deficiency of measurement on the north and west sides of the townships. For convenience the exterior lines on which subdivisions are based are called the governing boundaries. In unusual cases the north and west boundaries may be employed to govern the subdivision of a township, and in extreme cases an irregular township may be without even a single governing boundary.

## INITIAL POINTS.

**138.** Initial points from which the lines of the public surveys are to be extended will be established whenever necessary, under such special instructions as may be prescribed in each case by the Commissioner of the General Land Office. The initial points are to be selected with a view to their control of extensive agricultural areas within reasonable geographical limitations. Upon the establishment of an initial point, the position of the point in latitude and longitude is to be determined by accurate field astronomical methods.

During the period since the organization of the system of rectangular surveys numbered and locally named principal meridians and base lines have been established as shown by the accompanying tabular exhibit. These bases and meridians may be found by examining the large wall map of the United States published by the General Land Office; they are also shown upon the various official State maps, and upon a special map entitled "United States, Showing Principal Meridians, Base Lines and Areas Governed Thereby."

**139.** The latitudes and longitudes given in the following table are based upon the best obtainable information, but in some cases the values shown are only approximately correct owing to the fact that many of the initial points were fixed in position and the surveys therefrom largely completed before the same importance was attached to the matter of accurate latitudes and longitudes as at the present time. It may also be noted, by way of explanation, that present-day facilities for accurate field astronomical determinations were not available to the early surveyors. It is not expected that the values of the latitudes given in the table will be used as the basis of the calculation of the latitude of an unknown station, in lieu of a field determination thereof, except as an approximate value may satisfy all requirements. The coördinates of the earliest surveys in Ohio can not be conveniently tabulated, but they are shown upon the maps as stated above.

### PRINCIPAL MERIDIAN.

**140.** This line shall conform to the true meridian and will be extended from the initial monument, either north or south, or in both directions, as the conditions may require; regular quarter-section and section corners will be established alternately at intervals of 40 chains, and regular township corners at intervals of 480 chains; meander corners will be established at the intersection of the line with all meanderable bodies of water.

**141.** In the survey of the principal meridian and the other *standard lines* (base lines, standard parallels and guide meridians), hereinafter described, two independent sets of measurements will

## MERIDIANS AND BASE LINES OF THE UNITED STATES RECTANGULAR SURVEYS.

| Meridians. | Governing surveys (wholly or in part) in States of— | Longitude of principal meridians west from Greenwich. | | | Latitude of base lines north from Equator. | | |
|---|---|---|---|---|---|---|---|
| | | ° | ′ | ″ | ° | ′ | ″ |
| Black Hills | South Dakota | 104 | 03 | 00 | 44 | 00 | 00 |
| Boise | Idaho | 116 | 24 | 15 | 43 | 22 | 31 |
| Chickasaw | Mississippi | 89 | 15 | 00 | 34 | 59 | 00 |
| Choctaw | ....do | 90 | 14 | 45 | 31 | 54 | 40 |
| Cimarron | Oklahoma | 103 | 00 | 00 | 36 | 30 | 00 |
| Copper River | Alaska | 145 | 18 | 42 | 61 | 49 | 11 |
| Fairbanks | Alaska | 147 | 38 | 33 | 64 | 51 | 49 |
| Fifth Principal | Arkansas, Iowa, Minnesota, Missouri, North Dakota, and South Dakota. | 91 | 03 | 42 | 34 | 44 | 00 |
| First Principal | Ohio | 84 | 48 | 50 | 41 | 00 | 00 |
| Fourth Principal | Illinois | 90 | 28 | 45 | 40 | 00 | 30 |
| Do | Minnesota and Wisconsin | 90 | 28 | 45 | 42 | 30 | 00 |
| Gila and Salt River | Arizona | 112 | 17 | 25 | 33 | 22 | 40 |
| Humboldt | California | 124 | 08 | 00 | 40 | 25 | 12 |
| Huntsville | Alabama | 86 | 34 | 45 | 35 | 00 | 00 |
| Indian | Oklahoma | 97 | 14 | 30 | 34 | 30 | 00 |
| Louisiana | Louisiana | 92 | 24 | 15 | 31 | 00 | 00 |
| Michigan | Michigan | 84 | 22 | 24 | 42 | 26 | 30 |
| Mount Diablo | California and Nevada | 121 | 54 | 48 | 37 | 51 | 30 |
| Navajo | Arizona and New Mexico | 108 | 32 | 45 | 35 | 45 | 00 |
| New Mexico Principal | Colorado and New Mexico | 106 | 53 | 40 | 34 | 15 | 25 |
| Principal | Montana | 111 | 38 | 50 | 45 | 46 | 48 |
| Salt Lake | Utah | 111 | 54 | 00 | 40 | 46 | 04 |
| San Bernardino | California | 116 | 56 | 15 | 34 | 07 | 10 |
| Second Principal | Illinois and Indiana | 86 | 28 | 00 | 38 | 28 | 20 |
| Seward | Alaska | 149 | 21 | 53 | 60 | 07 | 26 |
| Sixth Principal | Colorado, Kansas, Nebraska, South Dakota, and Wyoming. | 97 | 23 | 00 | 40 | 00 | 00 |
| St. Helena | Louisiana | 91 | 09 | 15 | 31 | 00 | 00 |
| St. Stephens | Alabama and Mississippi | 88 | 02 | 00 | 31 | 00 | 00 |
| Tallahassee | Florida | 84 | 16 | 42 | 30 | 28 | 00 |
| Third Principal | Illinois | 89 | 10 | 15 | 38 | 28 | 20 |
| Uintah | Utah | 109 | 57 | 30 | 40 | 26 | 20 |
| Ute | Colorado | 108 | 33 | 20 | 39 | 06 | 40 |
| Washington | Mississippi | 91 | 09 | 15 | 31 | 00 | 00 |
| Willamette | Oregon and Washington | 122 | 44 | 20 | 45 | 31 | 00 |
| Wind River | Wyoming | 108 | 48 | 40 | 43 | 01 | 20 |

be employed, unless subdivisional closings thereon are provided in the same assignment with the standard line, in which case the closings will furnish a satisfactory verification of the length of the lines thus surveyed. Where such closings are not to be made during the progress of the same survey, the proper supervising officer will provide suitable instructions for the employment of a second set of chainmen, or for the duplication of the measurement by the one set of chainmen. In either case, where two independent sets of measurements are employed, the distance to the mean point, and the difference between the measurements to each corner established, will be shown in the field notes; a form of record is given in the specimen field notes.

**142.** Should the difference between the two sets of measurements of any standard line, as above provided, exceed 20 links per 80 chains, it is required that the line be remeasured to reduce the difference, the final measurement of the line only to be shown in the field notes. Should the successive independent tests of the alinement of any standard line, or the average tests of the solar attachment employed in the projection thereof, indicate that the line has deflected from the true cardinal course to exceed 3′ 00″, the necessary corrections will be made to reduce the deviation in azimuth, the field notes of the true line only being shown. Every reasonable effort will be exercised to insure the accuracy of both the alinement and the measurement of the standard lines, and the stated discrepancies are the maximum that will be allowed in new surveys; corrective steps will be required where the differences are beyond the maximum.

### BASE LINE.

**143.** From the initial monument the base line will be extended east and west on a true parallel of latitude; upon the true line standard quarter-section and section corners will be established alternately at intervals of 40 chains, and standard township corners at intervals of 480 chains; meander corners will be established at the intersection of the line with all meanderable bodies of water.

The manner of making the measurement of the base line and the accuracy of both the alinement and measurement will be the same as required in the survey of the principal meridian. Any one of the methods heretofore set forth for the determination of the alinement of the true latitude curve may be used as existing conditions may require and the detailed process will be fully stated in the field notes.

### STANDARD PARALLELS.

**144.** Standard parallels, which are also called correction lines, are extended east and west from the principal meridian, at intervals of 24 miles north and south of the base line, in the manner prescribed for the survey of the base line.

Fig. 16.

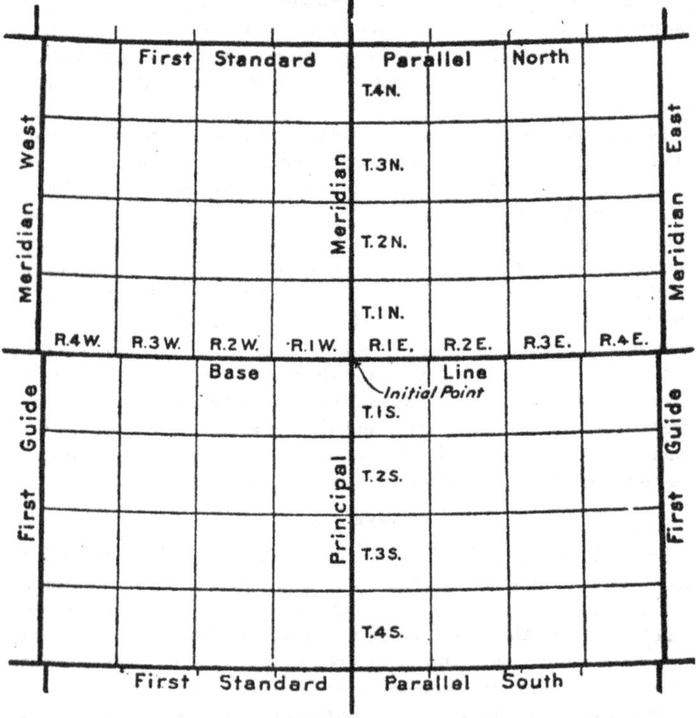


Illustrating the survey of quadrangles each embracing 16 townships bounded by standard lines, and showing the coördinate system of numbering the townships

**145.** Where standard parallels have been placed at intervals of 30 or 36 miles, under practice then permissible, and present conditions require additional standard lines from which to initiate new, or upon which to close the extension of old surveys, an intermediate correction line should be established to which a local name may be given, e. g., "Fifth Auxiliary Standard Parallel North," or "Cedar Creek Correction Line," etc., and the same will be run, in all respects, like a regular standard parallel.

### GUIDE MERIDIANS.

**146.** Guide meridians are extended north from the base line, or standard parallels, at intervals of 24 miles east and west from the principal meridian, in the manner prescribed for running the principal meridian. Under all conditions the guide meridians will be terminated at the points of their intersections with the standard parallels; the guide meridian is to be projected on the true meridian and the fractional measurement is to be placed in the last half mile. At the true point of intersection of the guide meridian with the standard parallel a closing township corner is to be established; the parallel will be retraced between the first standard corners east and west of the point for the closing corner, in order to determine the exact alinement of the line closed upon, and the distance will be measured and recorded to the nearest corner on said standard parallel.

**147.** When existing conditions require that such guide meridians shall be run south from the base or correction lines, they will be initiated at the theoretical point for the closing corner of the guide meridian, which will be calculated on the basis of the survey of the line from south to north initiated at the proper standard township corner. At the theoretical point of intersection a closing township corner will be established.

**148.** Where guide meridians have been placed at intervals exceeding the distance of 24 miles, and new governing lines are required in order to limit the errors of the old or to control new surveys, a new guide meridian will be established, and a local name may be assigned to the same, e. g., "Twelfth Auxiliary Guide Meridian West," or "Grass Valley Guide Meridian," etc. These auxiliary guide meridians will be surveyed in all respects like regular guide meridians.

**149.** The above scheme covers the controlling lines contemplated under the rectangular system, and results regularly in the survey of quadrangles bounded on the north and south by true parallels of latitude, and on the east and west by true meridians, 24 miles apart. One exception may now be noted which will be found to depart from former practice, that is, where a guide meridian is carried forward at a time when uncertainty exists as to how the exterior and subdivisional surveys to the east may close upon it, the corners upon the same will be marked only for the surveys to the west.

### TOWNSHIP EXTERIORS.

#### REGULAR ORDER.

**150.** The controlling factors to be recognized in the establishment of new township boundary lines are found in the relation of these lines to the new subdivisional surveys which are to be executed. The south and east boundaries are normally the governing lines of the subdivisional surveys. Defective conditions which may be found in previously established exteriors can not be eliminated where subdivisional lines have been initiated from or closed upon an old boundary, but the errors of the former surveys are not to be incorporated into the new, and where the previously established south and east boundaries can not on that account be used to govern the subdivision of the adjoining township, other controlling lines known as the sectional correction line and the sectional guide meridian, hereinafter described, will be employed as expedient. A new meridional township exterior is normally the governing boundary of the township to the west, and a new latitudinal township exterior is normally the governing boundary of the township to the north; any new boundary should therefore be established with full consideration for its control upon the subdivisional surveys thereafter to be executed.

**151.** Whenever practicable the township exteriors will be surveyed successively through a quadrangle in ranges of townships, beginning with the townships on the south. The meridional boundaries of the townships will have precedence in the order of survey and will be run from south to north on true meridians; quarter-section and section corners will be established alternately at intervals of 40 chains, and meander corners at the intersection of the line with all meanderable bodies of water; a temporary township corner will be set at a distance of 480 chains, pending a determination of the controlling factor upon which its final position will be governed, whereupon the temporary point will be replaced by a permanent corner in proper latitudinal position. The latitudinal township boundary will be run first as a random line, setting temporary corners, on a cardinal course, from the old toward the new meridional boundary, and corrected back on a true line if ideal conditions are

Fig. 17.

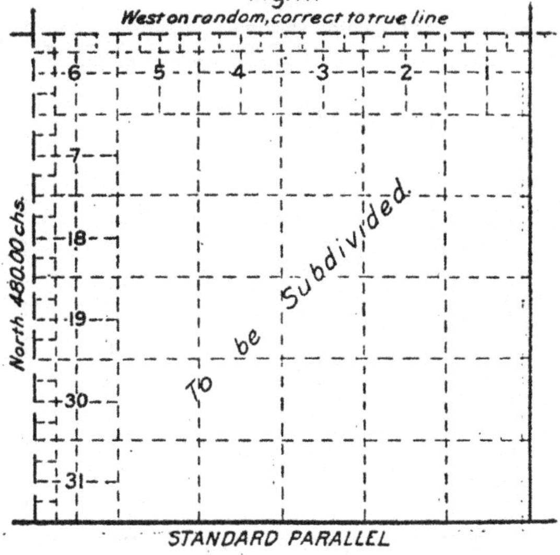
West on random, correct to true line
6
5
4
3
2
7
18
19
30
31
North 480.00 chs.
To be Subdivided.
STANDARD PARALLEL

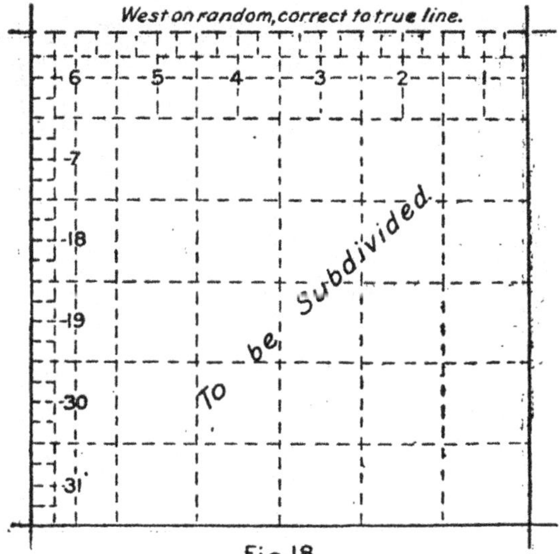
West on random, correct to true line.
6
5
4
3
2
7
18
19
30
31
To be Subdivided.

Fig. 18.

Fig. 19.

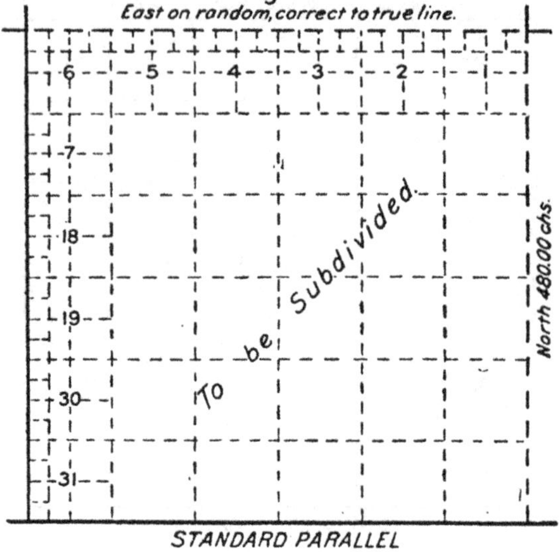
East on random, correct to true line.
6
5
4
3
2
7
18
19
30
31
To be Subdivided.
North 480.00 chs.
STANDARD PARALLEL

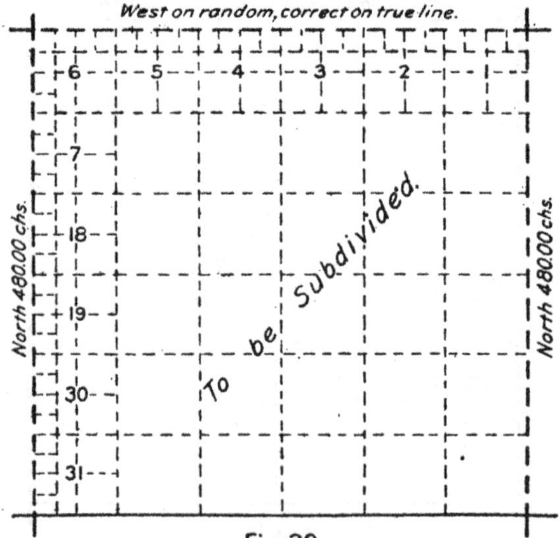
West on random, correct on true line.
6
5
4
3
2
7
18
19
30
31
To be Subdivided.
North 480.00 chs.
North 480.00 chs.

Fig. 20

Fig. 21.

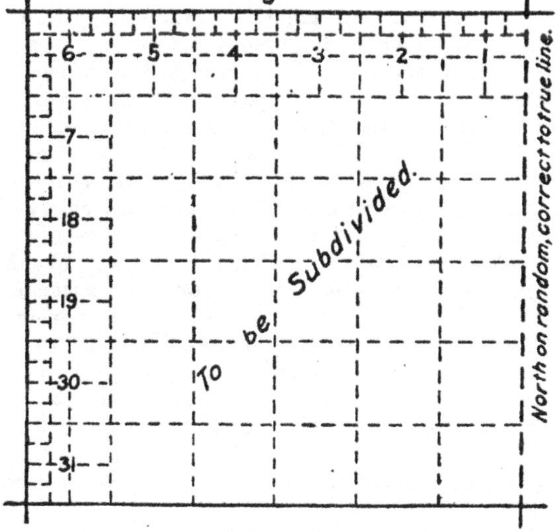


STANDARD PARALLEL

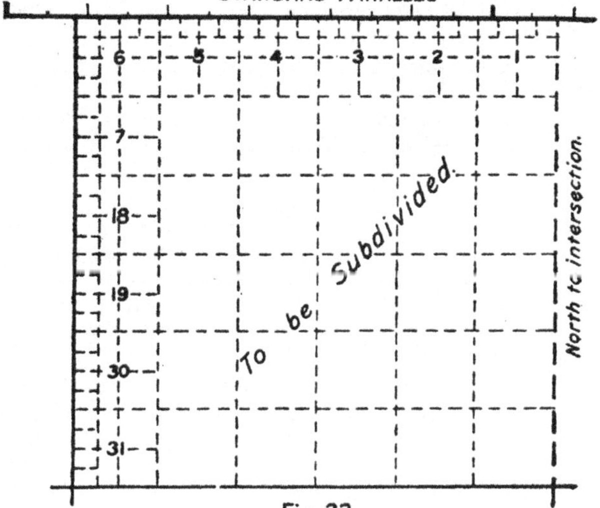


Fig. 22.

found to obtain. Where both meridional boundaries are new lines or where both have been previously established, the random latitudinal boundary will be run from east to west. In either case, if defective conditions are not encountered, the random line will be corrected back on a true line, upon which will be established regular quarter-section and section corners at intervals of 40 chains, alternately, counting from the east, and meander corners at the intersection of the true line with all meanderable bodies of water. The bearing of the true line will be calculated on the basis of the falling of the random, and the fractional measurement will be placed in the west half mile. A meridional township exterior will be terminated at the point of its intersection with a standard parallel, placing the excess or deficiency in measurement in the northernmost half mile. At the point of intersection of the meridional boundary with a standard parallel a closing township corner will be established; the parallel will be retraced between the first standard corners east and west of the point for the closing corner, in order to determine the exact alinement of the line closed upon, and the distance will be measured and recorded to the nearest corner on said standard parallel.

**152.** In order to complete the exteriors of a township it will often remain to establish a meridional boundary between previously established township corners; such boundaries will be run from south to north on random lines, with temporary corners set at intervals of 40 chains, and, if defective conditions are not encountered, the random will be corrected to a true line; by this plan the excess or deficiency of measurement will be placed in the north half mile, as required by law, and double sets of corners will be avoided where unnecessary.

**153.** The temporary points on any random exterior will be replaced by permanent corners, in proper position, when the final true line adjustments for the latter have been fully determined; the true line will be properly blazed through timber, and distances to important items of topography will be adjusted to correct true line measurements.

**154.** The field notes will embrace a full and complete record of the manner in which the township exteriors are run and established. The notes will show how the alinement of the random latitudinal curve was determined, the direction of the projection, the amount of the falling north or south of the objective township corner, and the calculated return course or true line.

### IRREGULAR ORDER AND PARTIAL SURVEYS.

**155.** As the remaining unsurveyed public lands are found to contain less and less extensive areas surveyable under the law it becomes necessary to depart from the ideal procedure in order more directly to reach the areas authorized for survey. The many possible combinations are entirely too numerous to state in detail, but where an irregular order appears to be necessary such departure from the ideal order of survey will be specifically outlined in the written special instructions. Such departure should always be based on the principle of accomplishing, by whatever plan, the same relation of one township boundary to another as would have resulted from regular establishment under ideal conditions.

In authorizing surveys to be executed it will not usually be provided that exteriors are to be carried forward until the township is to be subdivided; thus where causes operate to prevent the establishment of the boundaries in full it is not imperative that the survey of the exterior lines be completed; under such conditions it may be found necessary to run section lines as offsets to township exteriors and such section lines will be run either on cardinal courses or parallel to the governing boundaries of such townships, or even established when subdividing, as existing conditions may require.

### GENERAL EXCEPTIONS.

**156.** The above rules accord with former practice, except that in certain instances the random latitudinal boundaries will be run from west to east, instead of invariably from east to west, as heretofore required. It is also deemed advisable to incorporate other exceptions which will lessen the difficulties of subdivisional surveys frequently experienced in the past.

It is especially desirable that the alinement of a new latitudinal boundary (which becomes the governing south boundary of the township to the north) shall not depart more than 14′ from the true cardinal course; therefore the random line, run upon the cardinal course, may be made the true line where the falling would require a correction exceeding 14′ of arc. Where the random latitudinal boundary thus closes on a *new* meridional exterior the temporary township corner may be adjusted to the latitude of the opposite township corner; but where both meridional boundaries have been previously surveyed a closing township corner will be established at the point of intersection of the random latitudinal line with the

Fig. 23.

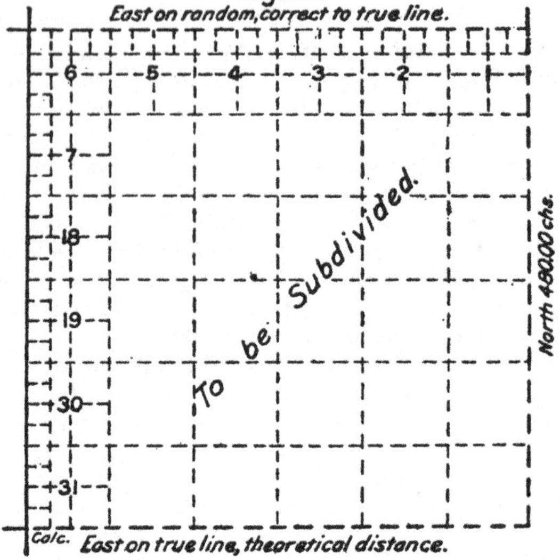
East on random, correct to true line.
6
5
4
3
2
7
18
19
30
31
To be Subdivided.
North 480.00 chs.
Calc.
East on true line, theoretical distance.

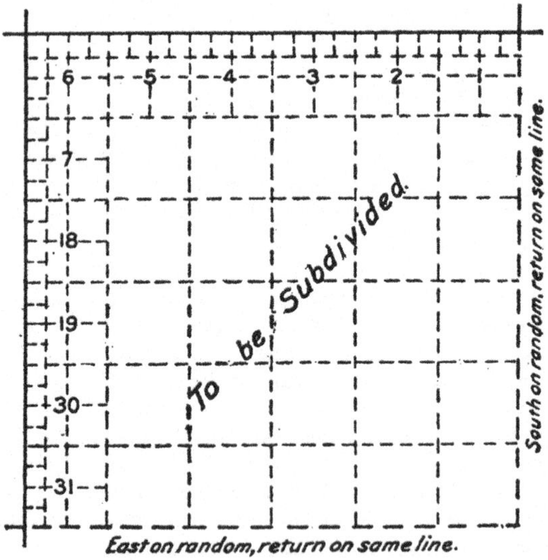
6
5
4
3
2
7
18
19
30
31
To be Subdivided.
South on random, return on same line.
East on random, return on same line.

Fig. 24.

Fig. 25.

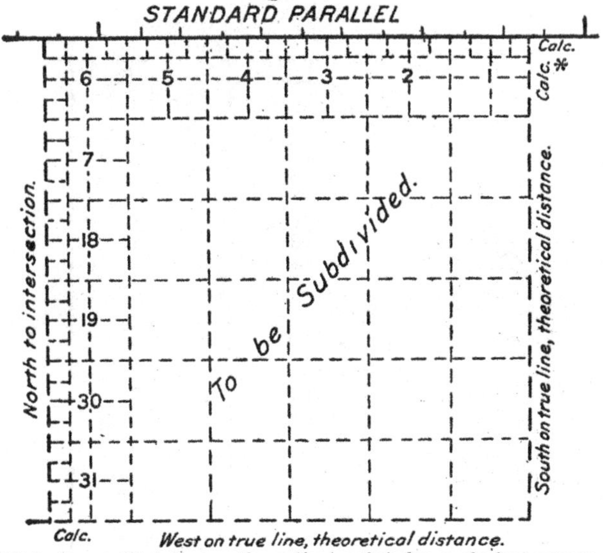


* Exteriors initiated at a theoretical point for a closing corner.

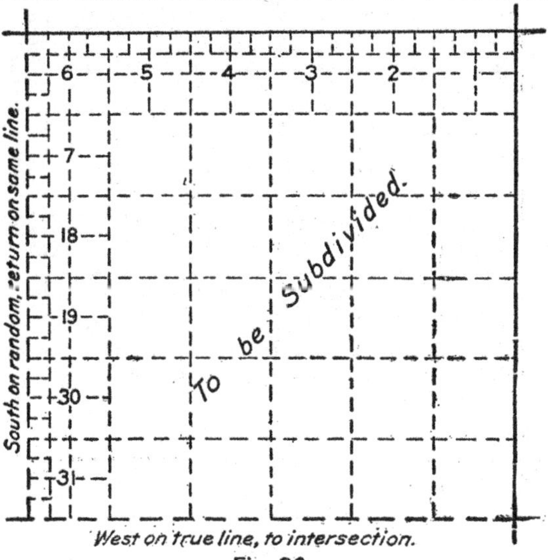


Fig. 26.

Fig. 27.

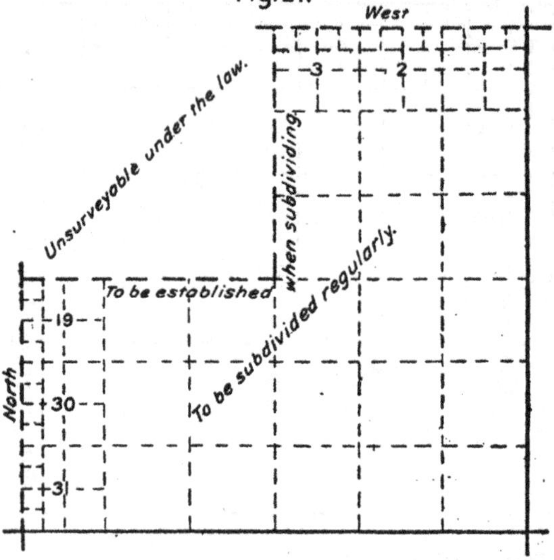
West
3
2
Unsurveyable under the law.
when subdividing
To be established
To be subdivided regularly.
North
19
30
31

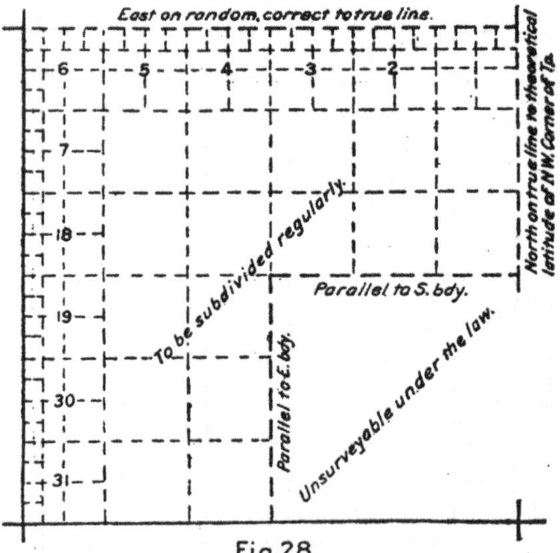
East on random, correct to true line.
6
5
4
3
2
7
18
19
30
31
To be subdivided regularly.
Parallel to S. bdy.
Parallel to E. bdy.
Unsurveyable under the law.
North on true line to theoretical latitude of N.W. Corner of Tp.

Fig. 28.

meridional boundary, or its projection to the north or south as the case may be. Likewise, where a meridional boundary is run as a random, the random will be made the true line if the adjustment for falling plus the usual correction to secure parallelism of the meridi-

Fig. 29.

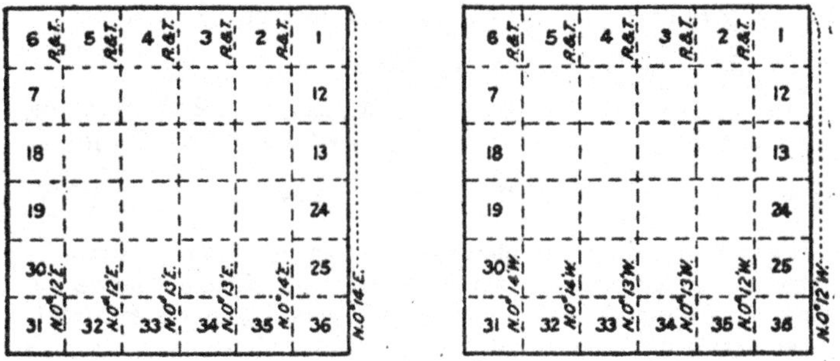

Latitude 25° N.

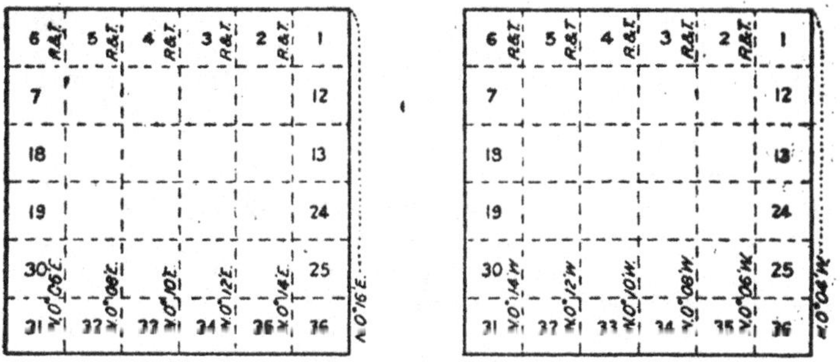

Latitude 70° N.

Illustrating the adjustment in the direction of the meridional lines of a subdivisional survey on account of convergency of meridians, also the 14′ limit of the rectangular "safety zone."

onal subdivisional lines (on account of convergency of meridians) would result in calculated bearings (in the northernmost miles of the latter lines) in excess of 14′ from cardinal. This margin for the alinement of the random and true meridional lines of the subdivisional survey calls for a governing east boundary whose bearing will fall within certain extremes suited to the latitude of the township, as for example (see second part of Table 2, Standard Field Tables):

*Latitude 25° N.*

| | | | |
|---|---|---|---|
| 1st Mi. Mer. Subdv. | N. 0° 14′ E. | 5th Mi. Mer. Subdv. | N. 0° 14′ W. |
| Corr. for Conv. | +00 | Corr. for Conv. | −02 |
| E. bdy. may be | N. 0° 14′ E. | E. bdy. may be | N. 0° 12′ W. |

*Latitude 70° N.*

| | | | |
|---|---|---|---|
| 1st Mi. Mer. Subdv. | N. 0° 14′ E. | 5th Mi. Mer. Subdv. | N. 0° 14′ W. |
| Corr. for Conv. | +02 | Corr. for Conv. | −10 |
| E. bdy. may be | N. 0° 16′ E. | E. bdy. may be | N. 0° 04′ W. |

It will be noted that the above text in reference to the 14′ limit for exteriors applies only to the establishment of new boundaries. A previously established boundary every part of which is within 21′ of cardinal will not be considered defective in alinement. Even in the case of new exteriors, where the surveyor who establishes such line is also to subdivide the township of which such exterior is a governing boundary, the margin of 14′ may be exceeded to a limited extent if the surveyor is satisfied that existing conditions favor keeping within the 21′ limit in the subdivisional survey. Thus it will be seen that the purpose of the 14′ limit is merely to facilitate the establishment of all subdivisional lines within the prescribed definite limit of 21′ from cardinal.

**157.** Another general exception may be noted where uncertainty exists as to how unsurveyed exteriors and subdivisional lines will close upon the newly established boundaries, in which case the corners thereon may be marked only for the townships of which the new exteriors control the subdivisions.

### COMPLETION OF PARTIALLY SURVEYED EXTERIORS.

**158.** Where the end portions of a township exterior have been previously surveyed and closed upon, the fractional unsurveyed middle part will be completed by random and true line, without offset regardless of the deviation from cardinal; the fractional measurements will be placed as a general rule in the north and west half miles, thereby permitting the subdivisional lines to be extended as usual from the south to the north and from the east to the west. In the case of a fractional part of an exterior remaining unsurveyed at either end of the line, the boundary will be completed by random line, initiated at the previously established terminal monument, which will be projected on a cardinal course in the direction of the objective township corner. The random will be corrected to a true line where the calculated bearing of any subdivisional line, governed by such exterior, comes within 14′ from cardinal, and the

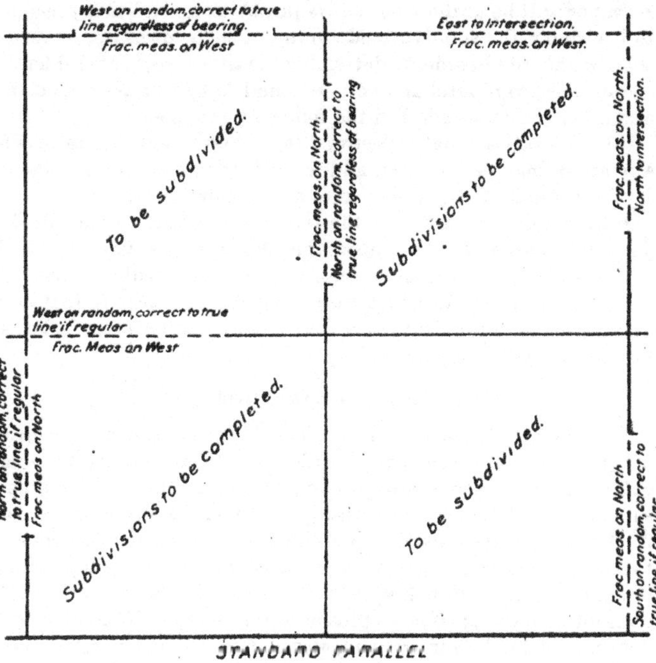


Fig. 30.

fractional measurement will be placed generally in the north or west half miles. However, should irregularity be developed, or in the absence of a previously established objective township corner, the partially surveyed exteriors will be completed on cardinal courses beginning as above; and in either case the fractional measurements will generally be placed in the north and west half miles.

### RETRACEMENTS BEFORE SUBDIVIDING.

**159.** If any part or all of the boundaries of a township which is to be subdivided have been previously surveyed, and the proper supervising officer has reason to question the accuracy of any portion of such exteriors, or the condition of the corner monuments thereon, the fact will be stated in the written special instructions, and the surveyor will be authorized and required, as a condition precedent to beginning the subdivisional survey of such township, to retrace such boundaries in order to determine the true alinement and lengths of the lines, to rebuild any corners found to be in a poor condition, and otherwise to accomplish the following purposes:

(*a*) To locate all material errors, (*b*) to test every line as to what alterations may be required, and (*c*) to determine all data necessary for the computation of the areas of all fractional lots.

**160.** All data obtained in the retracements will be embodied in the field notes and shown upon the plat of the survey, unless the retracement results are in substantial agreement with the record of the original survey, in which case a general statement to that effect may be made in the field notes, and the original record may be permitted to govern the data to be placed upon the plat.

### RECTANGULAR LIMITS.

**161.** Before approaching the subject of "subdivision of townships" it is necessary to consider the requirement of law relative to rectangular surveys, wherein the square mile, or section, is the unit of subdivision. The normal township will include 36 sections in all, 25 of which are returned as containing 640 acres each; 10 sections (on the north and west boundaries) each contain regular aliquot parts totaling 480 acres with 4 additional fractional lots in each section, each lot containing 40 acres plus or minus definite differences to be determined in the survey; and, section 6 containing regular aliquot parts totaling 360 acres with 7 additional fractional lots each containing 40 acres plus or minus certain definite differences to be determined in the survey, all as contemplated by law. The aforementioned aliquot parts of 640 acres may be termed "regular or legal subdivisions of a section," as a quarter section, a half-quarter section, or a quarter-quarter section, the legal minimum of which, for purposes of disposal under the general land laws, is 40 acres.

**162.** In the administration of the surveying laws it has been necessary to establish a definite relation between rectangularity

Fig. 31.

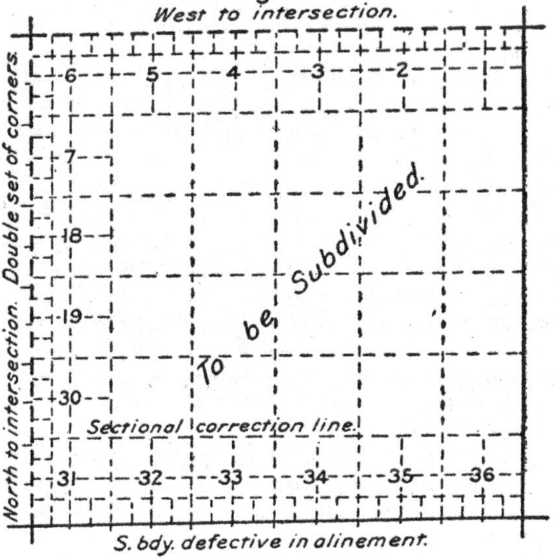
West to intersection.
North to intersection. Double set of corners.
6
5
4
3
2
7
18
19
30
31
32
33
34
35
36
To be Subdivided.
Sectional correction line.
S. bdy. defective in alinement.

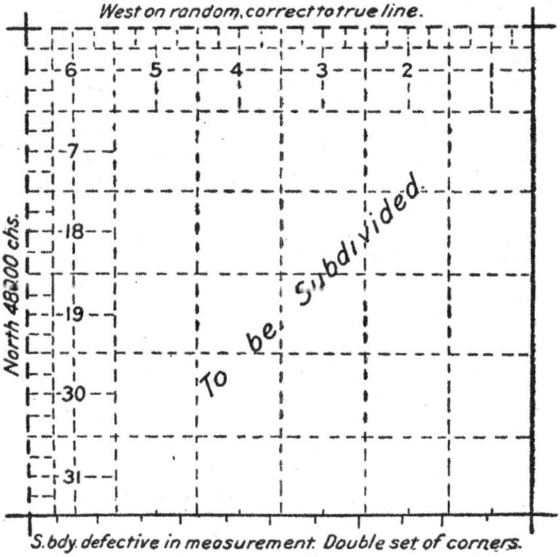
West on random, correct to true line.
North 48.00 chs.
6
5
4
3
2
7
18
19
30
31
To be Subdivided.
S. bdy. defective in measurement. Double set of corners.

Fig. 32.

(square miles of 640 acres, or aliquot parts thereof), as contemplated by law, and the resulting unit of subdivision consequent upon the practical application of surveying theory to the marking out of the lines on the earth's surface, wherein the ideal section is allowed to give way to one which may be termed "regular." Such relation, as applied to the boundaries of a section, has been placed at the following limits:

(*a*) For alinement, not to exceed 21′ from cardinal in any part; (*b*) for measurement, the distance between regular corners to be normal according to the plan of survey, with certain allowable adjustments not to exceed 25 links in 40 chains; and (*c*) for closure, not to exceed 50 links in either latitude or departure.

Township exteriors, or portions thereof, will be considered defective when they do not qualify within the above limits. It is also necessary, in order to subdivide a township regularly, to consider a fourth limit, as follows:

(*d*) For position, the corresponding section corners upon the opposite boundaries of the township to be so located that they may be connected by true lines which will not deviate more than 21′ from cardinal.

A previously established exterior will not be considered defective if the above limits are satisfied, and a *subdivisional survey* may proceed in safety if the rectangular limits (in such subdivisional survey) are not exceeded. On the other hand, if the conditions relating to the previously established *governing boundaries* are such that the rectangular limits have already been exceeded or that the danger point is likely to be reached at an early stage in the subdivisional survey, the necessary corrective steps will be taken before subdividing, as hereinafter described.

### RECTIFICATION OF DEFECTIVE EXTERIORS BEFORE SUBDIVIDING AND METHOD OF ESTABLISHING NEW GOVERNING BOUNDARIES WHERE THE PREVIOUSLY SURVEYED EXTERIORS ARE FOUND TO BE DEFECTIVE.

**163.** Where subdivisional lines have been initiated from or closed upon an exterior prior to the subdivision of one of the adjoining townships, its alinement can not legally be changed. A defective boundary not so closed upon may be obliterated, after connecting the old with the new monuments, whereupon a new boundary will be projected in accordance with regular methods. If a legal claim of any character such as mineral, forest-homestead, small-holding,

Fig. 33.

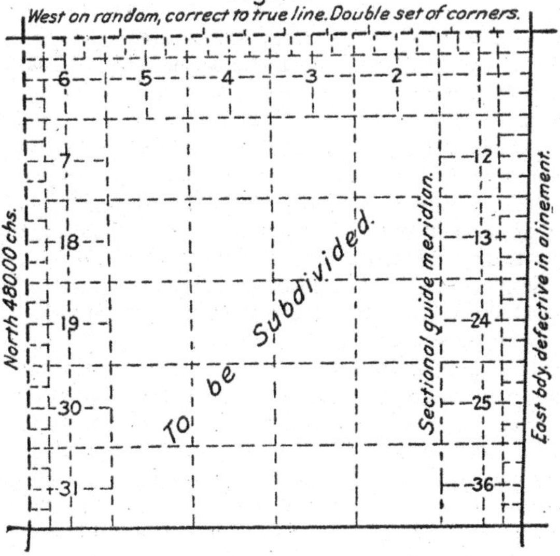


West to intersection.

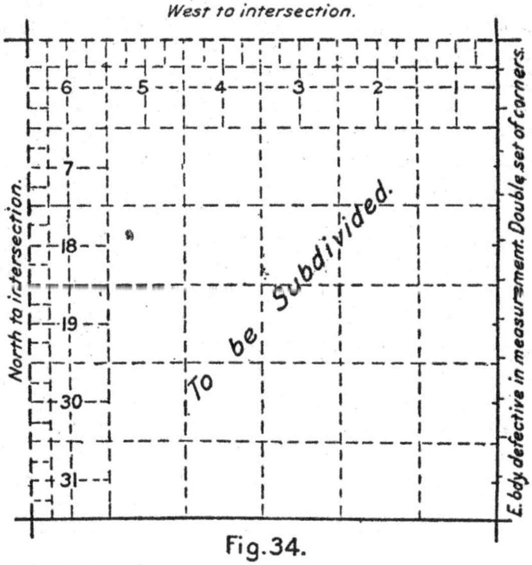


Fig. 34.

railroad or canal right-of-way, reservoir site, etc., has been connected with any corner on an exterior which may be subject to rectification, the fact will be specifically stated in the written special instructions, and in case such exterior is found to be defective the surveyor will accurately connect the old corner by course and distance with the new monument. Such old corners will not be destroyed, but the letters "W P" (signifying "witness point") will be distinctly added to the original markings. A complete record of the connection from the new to the old monument, a description of the latter and its accessories, and the new markings, will be included in the field notes, and the position of the old monument will be shown on the plat of the survey.

**164.** If a boundary is defective in measurement or position and is not subject to rectification, the location of the original corners will not be changed, but the marks thereon, and the marks upon or position of the accessories, may be appropriately altered to stand only for the sections of the previously established surveys. New corners to control the surveys of the adjoining township may then be established on the old line, but at regular distances of 40 and 80 chains. Where new corners are placed on an oblique exterior (one whose bearing departs more than 1° from cardinal) the same will be so located for measurement that the oblique distance multiplied by the cosine or sine of the bearing angle, as the case may be, will result in cardinal equivalents of 40 and 80 chains.

**165.** Where subdivisional lines have been initiated from or closed upon one side of a portion of a township boundary prior to the subdivision of the township on the opposite side, while upon the remaining portion of the same such conditions do not interfere, said remaining portion may be obliterated, if found defective, whereupon a new line will be projected in accordance with regular methods.

**166.**. The position of the new exteriors, or of new corners on defective township boundaries must be established by an actual rerunning of such lines; the data acquired in surveying subdivisional lines closing upon defective exteriors can not be accepted in lieu of such retracement or resurvey.

**167.** Instances will occur both in closing subdivisional surveys upon regular exteriors and in the retracement of defective boundaries not subject to rectification where it will be developed that the original monuments have become lost or obliterated, or where such

Fig. 35.

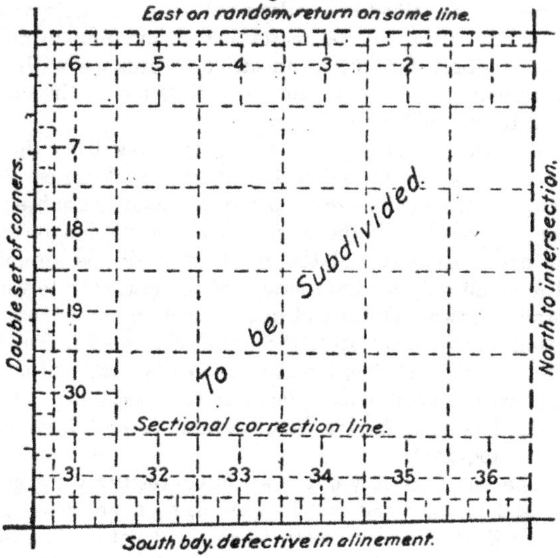
East on random, return on same line.
6
5
4
3
2
7
18
19
30
31
32
33
34
35
36
To be Subdivided.
Sectional correction line.
Double set of corners.
North to intersection.
South bdy. defective in alinement.

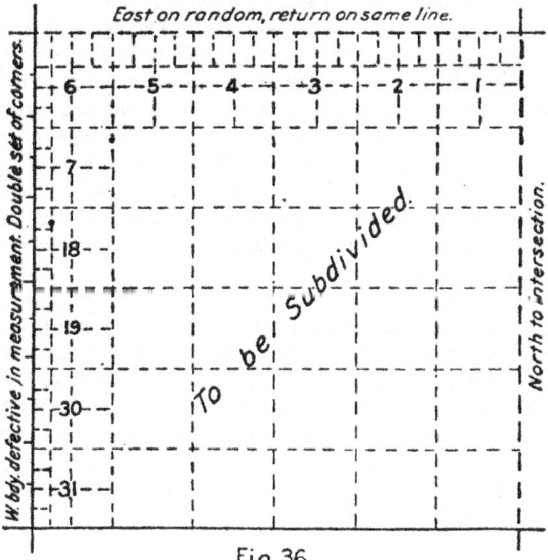
East on random, return on same line.
6
5
4
3
2
1
7
18
19
30
31
To be Subdivided.
W. bdy. defective in measurement. Double set of corners.
North to intersection.

Fig. 36.

corners may be identified in an advanced state of deterioration. All such exterior corners will be reestablished and remonumented in their correct original positions in strict accordance with the provisions of Chapters IV and V, and a complete record thereof will be embodied in the field notes.

**168.** The south boundary of a township is regularly the governing latitudinal boundary and will be used as such unless defective in alinement; if defective in measurement, and not subject to rectification, the position of the original corners will not be changed, but the marks thereon and the accessories will be appropriately altered to stand only for the sections of the township to the south; new corners of two sections and quarter-section corners common to the sections of the township to the north will be established at regular intervals of 40 chains, counting from the east, and the excess or deficiency in measurement placed in the west half mile. If the south boundary is defective in alinement, a sectional correction line will be required.

**169.** The east boundary of a township is regularly the governing meridional boundary and will be used as such unless defective in alinement; if defective in measurement, and not subject to rectification, the position of the original corners will not be changed, but the marks thereon and the accessories will be appropriately altered to stand only for the sections of the township to the east; new corners of two sections and quarter-section corners common to the sections of the township to the west will be established at regular intervals of 40 chains, counting from the south. If the east boundary is defective in alinement a sectional guide meridian will be required.

**170.** New west and north boundaries of a township become the governing meridional and latitudinal boundaries of the townships to the west and north, respectively, and are required to be properly established as such.

**171.** New east and south boundaries of a township become the closing meridional and latitudinal boundaries of the townships to the east and south, respectively, and where by peculiar necessity the ideal plan must be modified and doubt exists as to how unsurveyed lines may close upon same, the corners thereon may be established common only to the sections of the township of which the new lines are the governing boundaries. The corners appropriate to the sections upon the opposite side will be duly established as closing corners at the time of the survey of the subdivisional lines

Fig. 37.

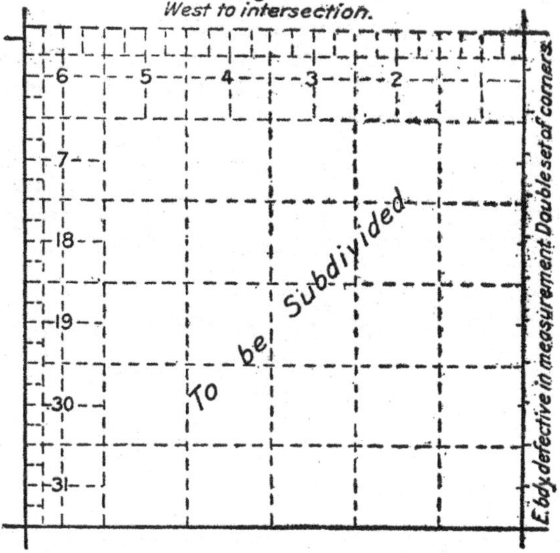
West to intersection.
6
5
4
3
2
7
18
19
30
31
To be Subdivided.
E. bdy. defective in measurement. Double set of corners.

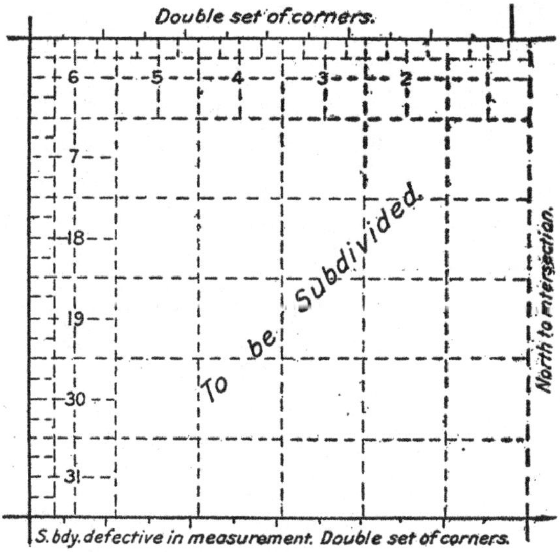
Double set of corners.
6
5
4
3
2
7
18
19
30
31
To be Subdivided.
North to intersection.
S. bdy. defective in measurement. Double set of corners.

Fig. 38.

of the adjoining townships if the original corners are then found to be defective in position, and where regular connections can be made the marks upon the original corners will be appropriately altered to corners of maximum control.

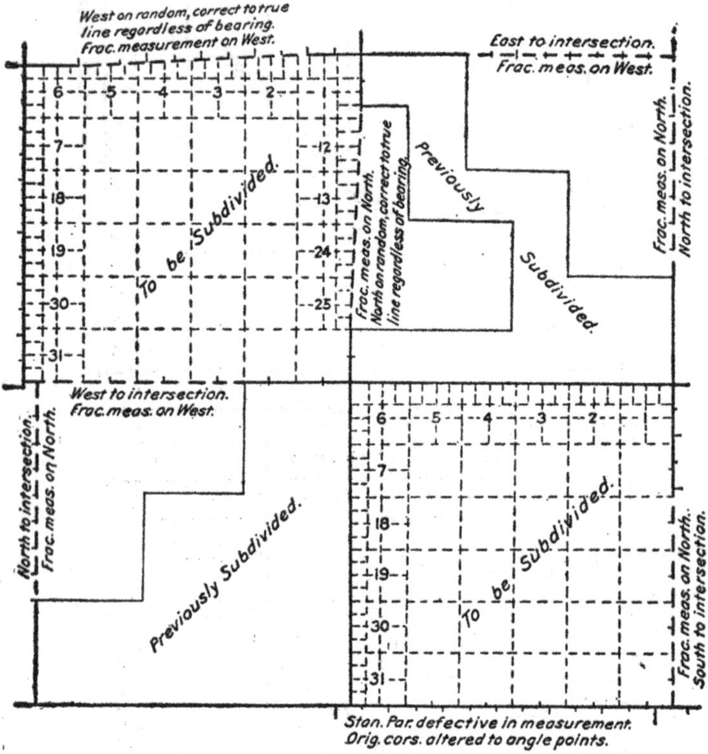


Fig. 39.

**172.** Where the previously established north or west boundaries are found to be defective in measurement or position and subdivisional surveys in the adjoining townships have been initiated upon the same, thereby preventing rectification, the marks upon the original corners will be appropriately altered to corners of two sections and quarter-section corners common only to the sections of the townships to the north or west, respectively. Closing section corners will be established when subdividing and the distance measured to an original corner; new quarter-section corners, common to the sections of the township which is being subdivided, will be placed on the old line at the mean distances between the closing section corners, or at 40 chains from one direction, depending upon the plan of the subdivision of the section. Where such previously established north and west boundaries are defective in alinement, but not in measurement or position, no changes are required, and the section lines of the township which is being subdivided will be connected regularly to the original corners; the resulting fractional measurements will be placed uniformly in the north and west half miles.

**173.** The diagrams which accompany the text illustrate the guiding principles involved in the method of establishing new governing boundaries where the previously surveyed exteriors are found to be defective. Each diagram illustrates a simple condition affecting one boundary only, and the examples are taken only from the regular order of procedure. Combinations of two or more of the simple defective conditions are best solved by an analysis of the complex problem into its several parts of simple defective conditions. The same statement is applicable to the solution of complex defective conditions encountered in the establishment of township exteriors under an irregular order of procedure. The surveyor will be expected to exercise skill and judgment in dealing with similar field problems, but where extraordinary conditions are encountered which will not admit of analysis and solution in harmony with the principles herein set forth he will report the facts to the proper supervising officer for his counsel.

## TABLES OF LATITUDES AND DEPARTURES AND CLOSING ERRORS.

**174.** Upon the completion of the survey of one or more township exteriors closing the figure of either a full or fractional township, a table of latitudes and departures and closing errors will be prepared, wherein due allowance for convergency of meridians will be introduced. The closing errors will furnish an immediate guide to the accuracy of the lines included in the table and, in case the limit of closure ($\frac{1}{640}$ of the perimeter, in either latitude or departure) is exceeded, will serve to show what additional retracements or other corrective steps may be necessary in order to perfect the survey before leaving the field. The table of latitudes and departures and closing errors, including every part of any closed figure embracing township exteriors, based upon final field determination after all necessary retracements and final true lines have been completed, will be incorporated in the field notes of the survey. The general subject of "limits of closure" will be amplified hereinafter.

## SUBDIVISION OF TOWNSHIPS.

### REGULAR BOUNDARIES.

**175.** The boundaries of a township will be considered within satisfactory governing limits from which to control the subdivisional survey when the calculated position of the latter lines may be theoretically projected from said boundaries without invading the danger zone in respect to rectangular limits as previously described. The danger zone has already been placed at theoretical bearings exceeding 14′ from cardinal, and the corresponding zone in respect to lengths of lines may be placed at theoretical adjustments exceeding 33 links per mile.

**176.** The direction of the east boundary may qualify anywhere within the governing limits set forth under the subject of "township exteriors," and where this boundary is broken in alinement, but otherwise within the governing limits, its mean course will be adopted when considering the control upon the direction of the meridional subdivisional lines.

**177.** The subdivision of a township may proceed in the normal order, where the above conditions are satisfied, as follows:

The meridional section lines will be initiated at the regularly established section corners on the south boundary of the township and will be run from south to north parallel to the governing east boundary, or, in case the east boundary is within limits, but has been found by retracement to be imperfect in alinement, the meridional section lines will be run parallel to the mean course of such east boundary. Regular quarter-section and section corners will be established alternately at intervals of 40 chains, as far as the northernmost interior section corner. The last miles of the meridional section lines will be continued as random lines, each successive line being run parallel to the true east boundary of the section to which it belongs; a temporary quarter-section corner will be set at 40 chains, the distances will be measured to the points of intersection of the random lines with the north boundary of the township, and the fallings of the random lines east or west of the objective section corners will be noted. The randoms will then be corrected to true lines by returning to accomplish the required markings between the section corners, including the permanent establishment of the quarter-section corners on the true lines at distances of 40 chains from the south, thus placing the fractional measurements in the north half miles. The bearings of the true lines will be calculated on the basis of the fallings of the randoms (see Table 3, Standard Field Tables). Where the north boundary of the township is a base line or standard parallel, the last miles of the meridional section lines will be continued as true lines parallel to the east boundary of the township, setting permanent quarter-section corners at 40 chains from the south and closing section corners at the points of intersection of the several lines with the base or standard or correction line, where the distances will be measured to the nearest corners on said line. The adjustment of the bearing of all meridional section lines on account of convergency of meridians has already been explained in Chapter II.

**178.** The latitudinal section lines, except in the west range of sections, will normally be run from west to east on random lines parallel to the south boundaries of the respective sections, setting temporary quarter-section corners at 40 chains; the distances will be measured to the points of intersection of the random lines with the north and south lines passing through the objective section corners, and the fallings of the random lines north or south of said corners will be noted. Each random will be corrected to a true line by returning to accomplish the required markings between the section corners, including the permanent establishment of quarter-section corners at the *mid-points* on the true lines. The bearings of the true lines will be calculated on the basis of the fallings of the randoms (see Table 3, Standard Field Tables). In the west range of sections the random latitudinal section lines will be run from east to west, parallel to the south boundaries of the respective sections, and on the true lines the permanent quarter-section corners will be established at 40 chains from the east, thus placing the fractional measurements in the west half miles.

**179.** Meander corners will be established at the points of intersection of the several true lines with all meanderable bodies of water.

**180.** The meridional section lines will have precedence in the order of execution, and these will be surveyed successively, beginning with the first meridional section line counting from the east. A meridional section line will not be continued beyond a section corner until after the connecting latitudinal section line has been surveyed, and in the case of the fifth meridional section line, both latitudinal section lines connecting east and west will be surveyed before continuing with the meridional line beyond a section corner. The successive meridional lines may be taken up at the convenience of the surveyor at any time in order as previously stated, but none will be carried beyond uncompleted sections to the east. The field notes will be compiled in ranges of sections beginning with the easternmost, and the west two ranges will be compiled by alternating with the adjoining east and west sections. The specimen field notes exemplify the usual order of survey and the prescribed method of arranging the field notes.

**181.** Thus, to recapitulate, the subdivisional survey will be commenced at the corner of sections 35 and 36, on the south boundary of the township, and the line between sections 35 and 36 will be run parallel to the east boundary of the township, or to the mean course thereof, if it is imperfect in alinement, but within limits, establishing the quarter-section corner at 40 chains, and at 80 chains, the corner of sections 25, 26, 35 and 36. From the last-named corner, a random line will be run eastward, without blazing, parallel to the south boundary of section 36, to its intersection with the east boundary of the township, placing at 40 chains from the point of beginning, a post for temporary quarter-section corner. If the random line intersects said township boundary exactly at the corner of sections 25 and 36, it will be blazed back and established as the true line, the permanent quarter-section corner being established thereon, midway between the initial and terminal section corners. If the random intersects said township boundary to the north or south of said corner, the falling will be carefully measured, and from the data thus obtained, the true return course will be calculated, and the true line blazed and established, and the position of the quarter-section corner determined, as directed above. The meridional section line will be continued on the same plan, likewise the successive latitudinal section lines except that each random will be run parallel to the true south boundary of the section to which it belongs. After having established the west and north boundaries of section 12, the line between sections 1 and 2 will be projected northward, on a random line, parallel to the east boundary of the township, or to its mean course, as the case may be, setting a post for temporary quarter-section corner at 40 chains, to its intersection with the north boundary of the township. If the random intersects said north boundary exactly at the corner of sections 1 and 2, it will be blazed back and established as the true line, the quarter-section corner being established permanently in its original temporary position, and the fractional measurement thrown into that portion of the line between the

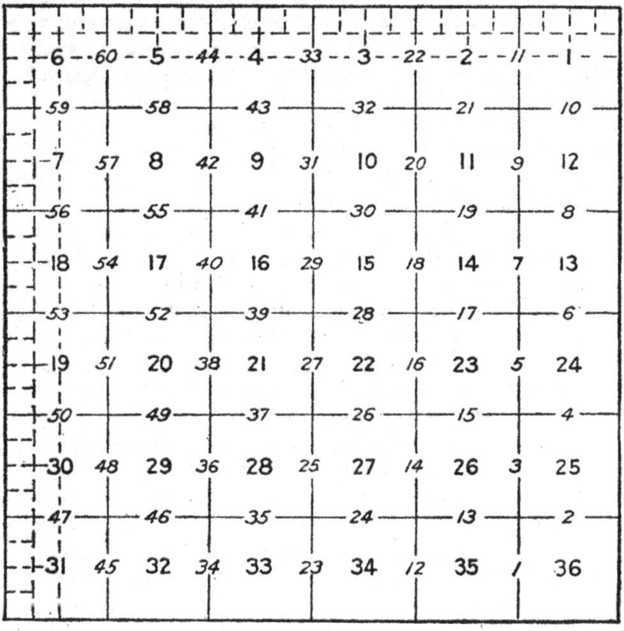


FIG. 40.—The numbers on the section lines indicate the normal order of subdivision and arrangement of the field notes.

permanent quarter-section corner and the north boundary of the township. If, however, said random intersects the north boundary of the township, to the east or west of the corner of sections 1 and 2, the falling will be carefully measured, and from the data thus obtained the true return course will be calculated and the true line established, the permanent quarter-section corner being placed upon the same at 40 chains from the initial corner of the random line, thereby throwing the fractional measurement in that portion lying between the quarter-section corner and the north boundary of the township. When the north boundary of a township is a base line or standard parallel, the line between sections 1 and 2 will be run as a true line parallel to the east boundary of the township, or to its mean course, as the case may be; the quarter-section corner will be placed at 40 chains, and a closing corner will be established at the point of intersection with such base or standard line; and in such case, the distance from said closing corner, to the nearest standard corner on such base or standard line, will be carefully measured and noted. The successive ranges of sections proceeding from east to west will be surveyed in the same manner; then after having established the west and north boundaries of section 32, a random line will be initiated at the corner of sections 29, 30, 31 and 32, which will be projected westward parallel to the south boundary of the township, setting a temporary quarter-section corner at 40 chains, to an intersection with the west boundary of the township, where the falling will be measured and the bearing of the true line calculated, whereupon the line between sections 30 and 31 will be permanently marked between the section corners, and the quarter-section corner thereon will be established at 40 chains from the east, thereby placing the fractional measurement in the west half mile as required by law. The survey of the west two ranges of sections will be continued on the same plan, and the random line between sections 6 and 7 will be run westward parallel to the true line between sections 7 and 18; the random will be corrected to a true line and the fractional measurement placed in the west half mile; finally the random line between sections 5 and 6 will be run northward parallel to the true line between sections 4 and 5; the random will be corrected to a true line and the fractional measurement placed in the north half mile.

It may well be noted again that the meridional section lines are surveyed as true lines for 5 miles, i. e., the lines are surveyed and permanently monumented in the first instance without later adjustment. Every means is placed at the disposal of the surveyor by which he is expected to accomplish accurate results, and the system of survey provides amply for the adjustment of all reasonable closing errors. Thus, a slight error in the alinement of the meridional section lines is taken up in the measurement of the latitudinal lines which, in order to come within the rectangular limit, must be within 50 links of 80 chains in length, except in the west range of sections where the convergency of the meridional lines is regularly provided for; the accumulated error in alinement for the 5 miles of true meridional line is taken up in the sixth mile, which is run random and true; here the true line must be within 21′ of cardinal in order to come within the rectangular limit. The slight, ordinary errors in the measurement of the meridional section lines are taken up by the adjustment of the bearings of the latitudinal section lines which, in order to come within the rectangular limit, must be within 21′ of cardinal; the accumulated error in measurement in running north is placed in the last fractional half mile; here the meridional distance will be checked by a calculated closing around the last section, and the latitudinal error must not exceed 50 links (or $\frac{1}{640}$) in order to come within the usual limits of closure. The accuracy of the subdivisional survey will everywhere be tested by the usual rules for limits of closure, hereinafter described. The surveyor should discriminate carefully between the limits for subdivision and limits of closure and note with due respect that whereas the latter may admit of differences as great as 50 links in any one section, the former are controlled by the limit of rectangularity and will be exceeded if the accumulative error is greater than 3½′ in alinement, or 8½ links per mile in measurement. The accumulative error must ever be guarded against and avoided, and the order of survey is arranged with a view to furnishing continuous checks upon the accuracy of all lines.

**182.** Any random subdivisional line may be run for distance only where the objective section corner is in sight, but the bearing will be recorded, and the usual rules for running random and true lines will be duly observed in every other respect. The random latitudinal section lines, except in the west range of sections, will normally be run from west to east, thus always closing upon a previously established section corner; but when under the exigencies of the field work, in order to economize the time of his party, the surveyor may elect to project the random from east to west (always parallel to the south boundary of the section), a temporary section corner (if the permanent corner has not already been established) will be set at 80 chains, and the true point for the section corner will be determined as usual at the 80-chain point on the meridional section line, whereupon the connection of the random latitudinal line and the permanent marking of the true line will be completed as regularly provided. Examples of the authorized rules for running subdivisional lines will be found in the specimen field notes.

### IRREGULAR BOUNDARIES.

**183.** Where either of the governing boundaries of a township is disqualified as a controlling line upon which to initiate a subdivisional survey, the necessary retracements and resurveys or alterations will be accomplished before subdividing as previously explained under the subject of township exteriors; thus may be assured every possible provision for a correct subdivisional survey except as either the south or the east boundary may be defective in alinement and not subject to rectification.

### SECTIONAL GUIDE MERIDIAN.

**184.** If the east boundary of the township is defective in alinement, and can not be rectified, and the north boundary is thus made defective in position, the first meridional section line will be projected on a true meridian to an intersection with the north boundary of the township where a closing section corner will be established and the distance measured to the nearest regular corner. The intermediate quarter-section and section corners will be established alternately at regular intervals of 40 chains, counting from the south, unless the south boundary of the township is itself defective in alinement. Where the north boundary is not defective in position (nor within the danger zone) with reference to the section corners on the south boundary (by reason of the errors in the alinement of the east boundary being compensating), the first meridional section line will be projected 5 miles as a true line on a bearing calculated to intersect the objective section corner on the north boundary, and the last mile will be run as a random line on the same course and corrected to a true line after the falling has been measured. The remaining meridional section lines will be run parallel to the one first established, in the usual manner, to closing section corners on the last mile or random and true as the case may be.

The fractional measurements of the latitudinal section lines in the first range of sections will be placed in the east half mile; elsewhere, unless the south boundary is defective in alinement, the latitudinal section lines will be run in the usual manner.

### SECTIONAL CORRECTION LINE.

**185.** If the south boundary of the township is defective in alinement, and can not be rectified, and the west boundary is thus made defective in position, a sectional correction line will be surveyed

Fig. 41.

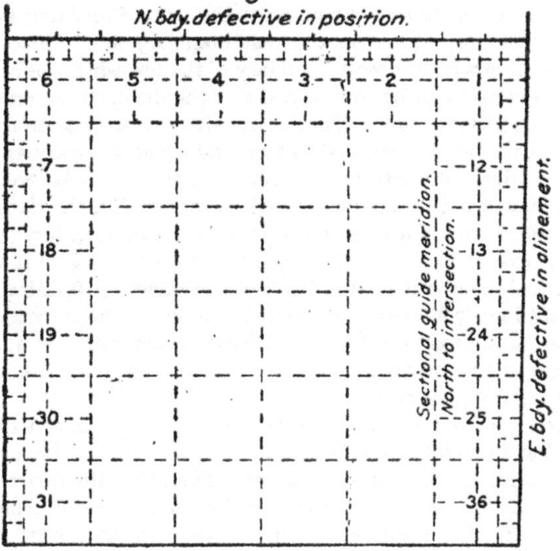
N. bdy. defective in position.
6
5
4
3
2
7
12
18
13
19
24
30
25
31
36
Sectional guide meridian.
North to intersection.
E. bdy. defective in alinement.

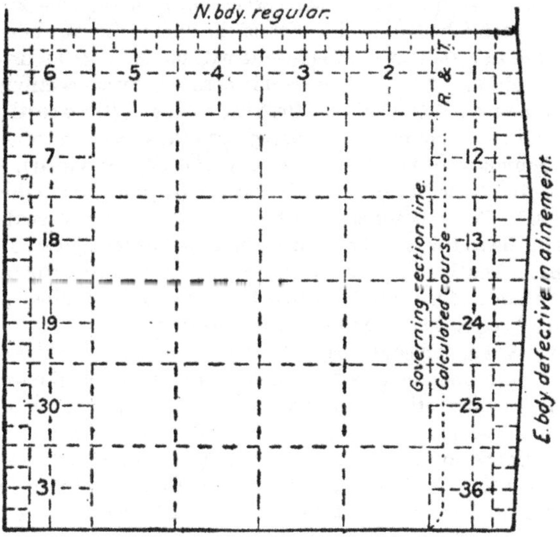
N. bdy. regular.
6
5
4
3
2
R. & T.
7
12
18
13
19
24
30
25
31
36
Governing section line.
Calculated course.
E. bdy. defective in alinement.

Fig. 42.

as a permanent line on a true latitudinal curve initiated at the first regular section corner on the east boundary and projected to an intersection with the west boundary of the township where a closing section corner will be established and the distance measured to the nearest regular corner. The intermediate quarter-section and section corners will be marked as temporary points at regular intervals of 40 chains, alternately, counting from the east. Where the west boundary is not defective in position (nor within the danger zone) with reference to the section corners on the east boundary (by reason of the errors in alinement of the south boundary being compensating), the first latitudinal section line will be projected 5 miles as a permanent line on a bearing calculated to intersect the objective section corner on the west boundary; temporary quarter-section and section corners will be marked at regular intervals of 40 chains, alternately, counting from the east.

The section corners on the sectional correction line will be established at the several points of intersection of the meridional section lines alined in the normal manner. Thereafter the quarter-section corners on the sectional correction line will be established at the usual mid-point positions except in the east and west ranges of sections. The quarter-section corner between sections 25 and 36 will be established at 40 chains from the west if the east boundary is defective in alinement; otherwise it will be fixed at the usual mid-point position. The quarter-section corner between sections 30 and 31 will be placed at 40 chains from the east, and if the sectional correction line has not been terminated at a closing section corner on the west boundary of the township (as previously provided), the line between sections 30 and 31 will be run random and true in the normal manner. The quarter-section corners on the meridional section lines in the south tier of sections will be permanently established at 40 chains south from the corners on the sectional correction line. The balance of the subdivisional lines will be continued from the sectional correction line in the usual manner.

**186.** Where the south part of the east boundary, or the east part of the south boundary, is regular, and the balance of the exterior is found to be defective in alinement and not subject to rectification, the subdivisional survey will be made regular as far as possible. The initial point for the sectional guide meridian, or for the sectional correction line, will be determined by existing conditions, and the subdivisional survey continued in harmony with the principles

Fig. 43.

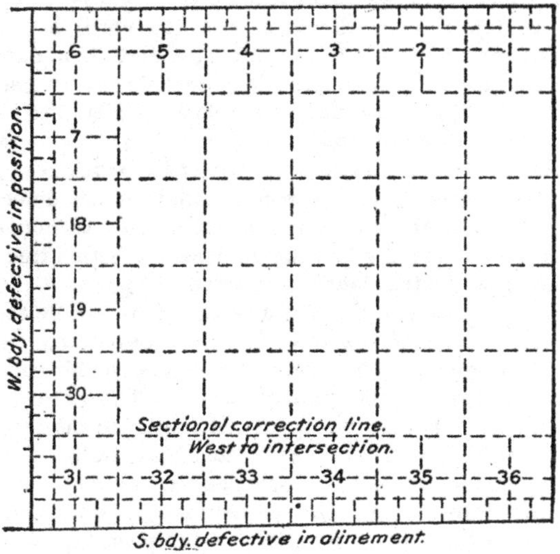

W. bdy. defective in position.
6
5
4
3
2
7
18
19
30
Sectional correction line.
West to intersection.
31
32
33
34
35
36
S. bdy. defective in alinement.


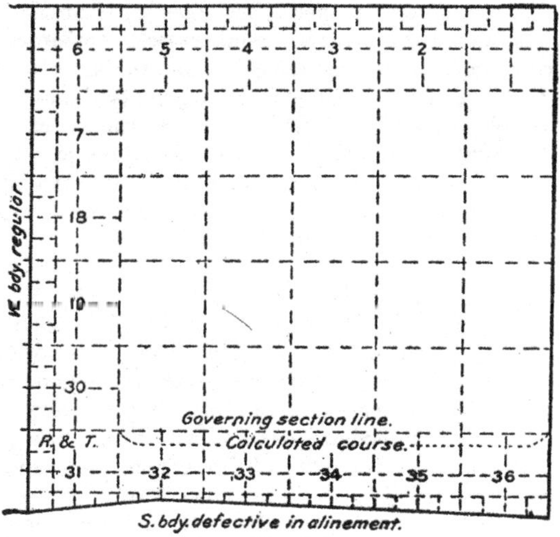

W. bdy. regular.
6
5
4
3
2
7
18
30
Governing section line.
R. & T.
Calculated course.
31
32
33
34
35
36
S. bdy. defective in alinement.


Fig. 44.

already outlined. Thus the first meridional section line would be continued as a sectional guide meridian if the north part of the east boundary is defective in alinement and the north boundary is thereby made defective in position, but if the north boundary is not defective in position (nor within the danger zone) the first meridional section line should be continued on a course calculated to intersect the objective section corner on the north boundary. The same principle would be observed if the west part of the south boundary is defective in alinement and the west boundary is not defective in position (nor within the danger zone), but if the west boundary is thus made defective in position the sectional correction line should be established on the true latitudinal curve.

Under the provisions of the above paragraph it will be seen that the maximum number of normal sections are to be secured where the condition of the governing boundaries warrants a combination of the several general plans of subdivisional surveys. The sections adjoining the east boundary may be considered regular to the full extent of their conformity with the usual rectangular limits, and where such agreement obtains the quarter-section corners on the latitudinal section lines will be placed at the normal mid-point position. The sections adjoining the south boundary of the township can not be considered regular unless the meridional lines are established at 80 chains in length, and the sections are otherwise in conformity with the usual rectangular limits; certain exceptions to this rigid requirement will be noted under the subject of "fragmentary subdivision."

**187.** The field notes of subdivisional surveys embracing either a sectional guide meridian, a sectional correction line, or other governing section line, will be compiled in the same regular order heretofore described, but appropriate explanatory remarks will be added indicative of the method and order of procedure.

### CLOSING SECTION LINES.

**188.** In the event of defective north or west boundaries, not subject to rectification, where the subdivisional lines can not be connected with the previously established exterior section corners, regularly by random and true lines not exceeding 21′ from cardinal and at the same time not deviating more than 21′ from a line parallel to the opposite (regular) boundary of the section, the normal positions of the randoms will be made the true lines; a closing section corner will then be established at the point of intersection of the

Fig. 45.

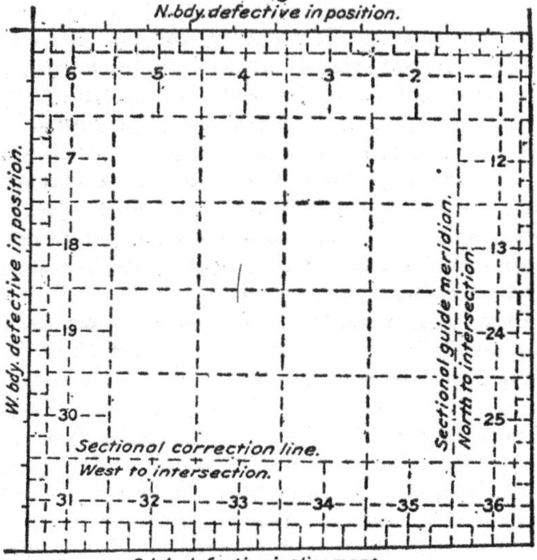
N. bdy. defective in position.
W. bdy. defective in position.
E. bdy. defective in alinement.
Sectional guide meridian.
North to intersection.
Sectional correction line.
West to intersection.
S. bdy. defective in alinement.
6
5
4
3
2
7
12
18
13
19
24
30
25
31
32
33
34
35
36

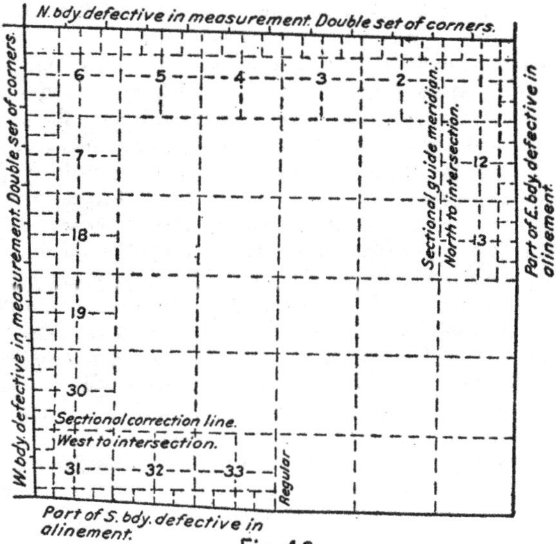
N. bdy defective in measurement. Double set of corners.
W. bdy. defective in measurement. Double set of corners.
Part of E. bdy. defective in alinement.
Sectional guide meridian.
North to intersection.
Sectional correction line.
West to intersection.
Regular
Part of S. bdy. defective in alinement.
6
5
4
3
2
7
12
18
13
19
30
31
32
33

Fig. 46.

section line with the original boundary, and the distance will be measured to the nearest original corner. The quarter-section corners on the closing section lines will be placed uniformly at 40 chains from the south or east as the case may be. If not already accomplished, the defective boundaries of the township will be retraced as may be necessary, and the marks upon the original corners appropriately altered as previously provided under the subject of rectification of defective exteriors, whereupon new quarter-section corners, common to the sections of the township which is being subdivided, will be established on the original defective boundaries at the mean distance between the closing section corners, or at 40 chains from one direction, depending upon the plan of the subdivision of the section to which a particular quarter-section corner belongs.

**189.** Corners of two sections on the governing south or east boundaries of a township will not be established as closing section corners, but at regular distances by measurement on said boundaries as already provided under the subject of rectification of defective exteriors before subdividing; thereafter the position of said corners will control the subdivisional survey.

**190.** Where a section is invaded by a State or reservation or grant boundary, or by a private claim of any description, such as mineral claims, forest-homestead claims, small-holding claims, etc., whose boundaries are at variance with the lines of legal subdivision, the distance on the township boundary or section line to the point of intersection with the irregular boundary will be carefully measured, likewise the exact bearing of the irregular boundary will be determined and the distance will be measured to the nearest corner on such irregular boundary. Where a private claim is located entirely within the limits of a section, a connection will be made from a regular corner on one of the boundaries of the section to a corner of the claim, and the bearing and length of the connecting line will be carefully determined. In the latter case a connecting traverse line will be recorded, if one is run, but it will also be reduced to the equivalent direct course and distance, all of which will be stated in the field notes, and the course and length of the direct connecting line will be shown upon the plat of the survey.

**191.** If a survey is to be concluded upon an irregular boundary at variance with the lines of legal subdivision, or if the survey is to be continued on a blank line to acquire a definite location upon the opposite irregular boundary, but without monumenting the rectangular survey between such irregular boundaries, a closing township or section corner, as the case may be, will be required at the point of intersection of the regular with the irregular line. On the other hand, if the survey is not to be so concluded, but is to be continued for the purpose of establishing a full complement of section and quarter-section corners for the control of the subdivision of a section so invaded by a private claim, no closing corner will be required.

**192.** In every case where a closing township or section corner is to be established upon a standard parallel, State, reservation, grant, or claim boundary, or upon an irregular section line or exterior, the line closed upon (if the latter was not established by the surveyor who runs the closing line, or if not already retraced by him), will be retraced between the first corners to the right and left of the point for the closing corner, in order to determine the exact alinement of the line closed upon, to the end that the closing corner may be established at the precise point of intersection of the two lines. The distance from the closing corner to the nearest corner on the line closed upon will always be measured and recorded.

## SUBDIVISION OF SECTIONS.

**193.** The acts of Congress approved February 11, 1805, and April 5, 1832, contain the fundamental provisions for the subdivision of sections into quarter sections and quarter-quarter sections; the principles recognized by law have already been stated in Chapter I. The sections are not subdivided in the field by the United States surveyors unless provision therefor is specifically mentioned in the written special instructions, but certain subdivision-of-section lines are always protracted upon the official plats, and the local surveyor who may be employed by entrymen to run said lines in the field is compelled to correlate the conditions as found upon the ground with those shown upon the approved plat. The United States surveyor is required to so establish the official monuments that a proper foundation is laid for the subdivision of the section, whereby the officially surveyed lines may be identified and the subdivision of the section controlled as contemplated by law.

**194.** The rectangular system provides for the unit of disposal under the general land laws, broadly, the quarter-quarter section of 40 acres, upon a plan in which the square mile, or section of 640 acres, is the unit of subdivision, while the unit of survey is the township of 36 sections. All agricultural entries are based upon descriptions in accordance with legal subdivisions shown upon the official plat. The plats are constructed in harmony with the official field notes returned by the surveyor. The land included in an entry is identified on the ground by fixed monuments established by the surveyor. A United States land patent grants to the entryman a title of ownership to a tract defined by certain fixed monuments on the ground and related by description and outline to the official plat. The function of the United States surveyor has been fulfilled when he has properly executed and monumented his survey and returned an official record thereof in the shape of complete detailed field notes and a plat. The function of the local surveyor begins when he is employed as an expert to identify the lands which have passed into private ownership; this may be a simple or a most complex problem, depending largely upon the condition of the original monuments as affected principally by the lapse of time since the execution of the official survey. The work of the local surveyor usually includes the subdivision of the section, already mentioned as the official unit of subdivision, into the fractional parts shown upon the approved plat. In this capacity the local surveyor is performing a function contemplated by law, and he can not properly serve his client or the public unless he is familiar with the legal requirements concerning the subdivision of sections. In the event that the original monuments have become lost the surveyor can not hope effectively to recover said corners without a full understanding of the record concerning their original establishment, nor can the surveyor hope legally to restore the same until he has mastered not only the principles observed in the execution of the original survey, but the principles upon which the courts having jurisdiction over such matters have based their rulings.

**195.** The General Land Office assumes no control or direction over the acts of local and county surveyors in the matters of subdivision of sections and reestablishment of lost corners of original surveys where the lands have passed into private ownership, nor will it issue instructions in such cases. It follows the general rule that disputes, arising from uncertain or erroneous location of corners, originally established by the United States, are to be settled by the proper local authorities or by amicable adjustment, and the office desires that the rules controlling the acts of its own surveying service be considered by all other surveyors as merely advisory and explanatory of the principles which should prevail in performing such duties.

The subject of restoration of lost corners will be treated in a later chapter, as the purpose here is to outline the principles concerning the subdivision of sections, which will be recognized alike by the General Land Office surveying service and by all local surveyors.

### SUBDIVISION BY PROTRACTION.

**196.** Upon the plat of all regular sections the boundaries of the quarter sections are shown by broken straight lines connecting the opposite quarter-section corners. The sections bordering the north and west boundaries of a normal township, excepting section 6, are further subdivided by protraction into parts containing two regular half-quarter sections and four lots, the latter containing the fractional areas resulting from the plan of subdivision of normal townships; the lines of the half-quarter sections are protracted from three points 20 chains distant from the line connecting the opposite quarter section corners, two of said distances counting on the opposite section lines and one counting on the line between the fractional quarter sections; the lines subdividing the fractional half-quarter sections into the fractional lots are protracted from mid-points on the opposite boundaries of the fractional quarter section. The two interior sixteenth-section corners on the boundaries of the fractional northwest quarter of section 6 are similarly fixed at points 20 chains distant north and west from the center of the section, from which points lines are protracted to corresponding points on the west and north boundaries of the section, resulting in subdivisions containing one regular quarter-quarter section and three fractional lots. The fractional lots herein described will be numbered in a regular series progressively from east to west or from north to south, in each section. As section 6 borders on both the north and west boundaries of the township, the fractional lots in the same will be numbered commencing with No. 1 in the northeast, thence progressively west to No. 4 in the northwest, and south to No. 7 in the southwest fractional quarter-quarter section.

Entrymen are allowed, under the law, to acquire title to any regular quarter-quarter section, but as such subdivisions are aliquot parts of quarter sections based upon mid-point protraction, it is not deemed necessary to indicate these lines upon the official plat.

Fig. 47.

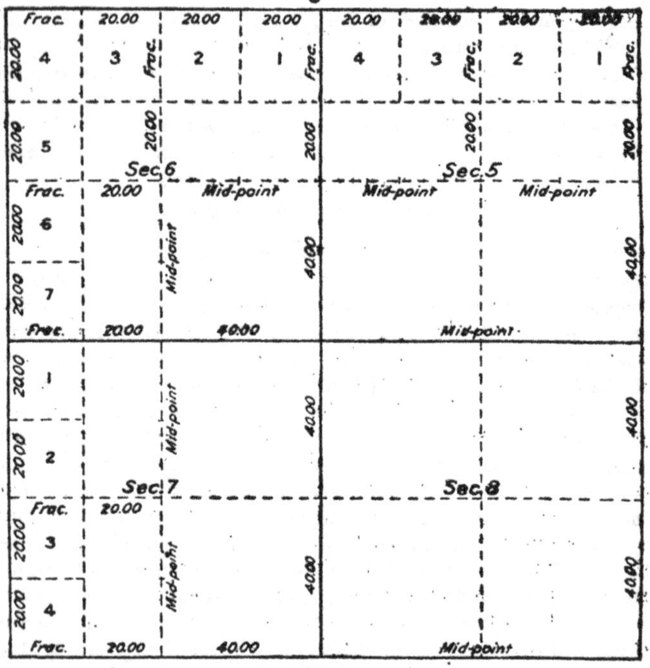


Showing normal subdivision of sections.

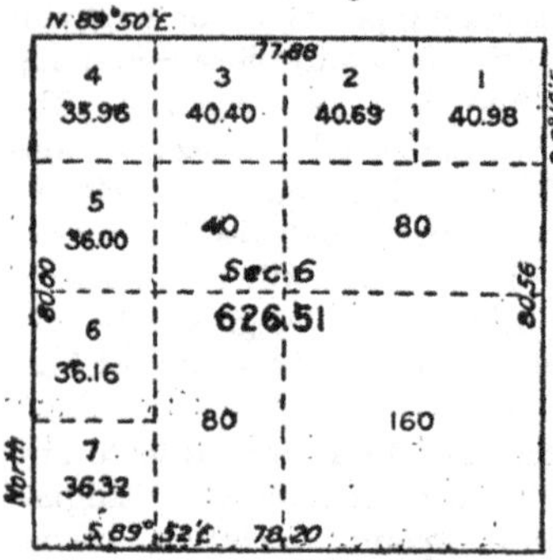


Showing areas.

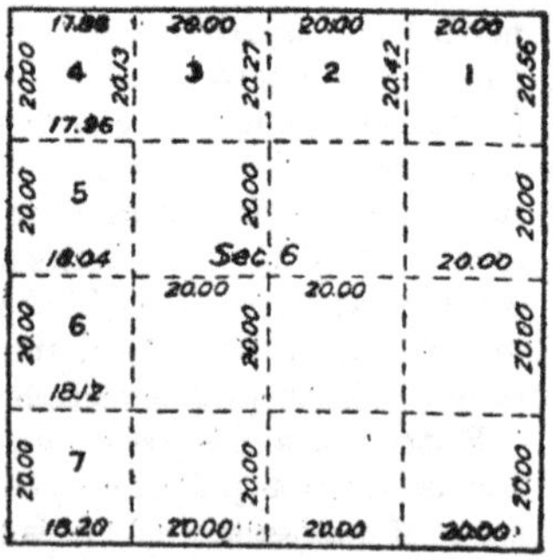


Showing calculated distances.

Examples of subdivision by protraction.

Fig. 48.

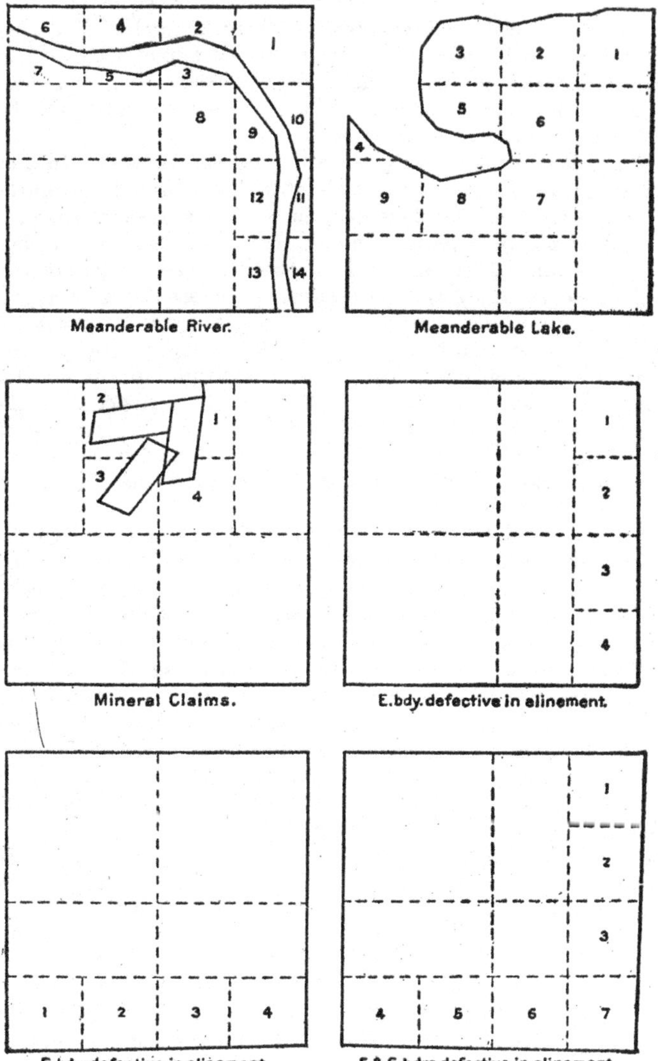


Examples of subdivision of fractional sections.

**197.** Sections which are invaded by meanderable bodies of water, or by approved claims at variance with the regular legal subdivisions, are subdivided by protraction into regular and fractional parts as may be necessary to form a suitable basis for the entry of the public lands remaining undisposed of, and to describe the latter separately and apart from the segregated areas.

The meander line of a body of water and the boundary lines of private claims are platted in accordance with lines run or connections made in the field; thereupon the sections so invaded are subdivided as nearly as possible in conformity with the uniform plan already outlined. The subdivision-of-section lines are terminated at the meander line or claim boundary, as the case may be, but the position of the subdivision-of-section lines is controlled precisely as though the section had been completed regularly. In the case of a section whose boundary lines are in part within the limits of a meanderable body of water, or within the boundaries of a private claim, the said fractional section lines are, for the purpose of uniformity, completed in theory, and the protracted position of the subdivision-of-section lines is controlled by the theoretical points so determined.

**198.** In the subdivision of fractional sections as many regular parts should be secured as possible, except to avoid thus creating poorly shaped fractional lots. Skill and judgment must be exercised to accomplish a subdivision which embraces simplicity of platting as well as a form to each and every lot that will prove to be equitable to the entryman. In the case of fractional lots along the north and west boundaries of a township, and in other similar cases where a lot has a full normal width of 20 chains in one direction, it is generally advisable to avoid areas of less than 10 or more than 50 acres, but in the instance of fractional lots along a meander line or other irregular broken boundary, where the width of the lot in both directions may be considerably less than 20 chains, resulting in tracts of more compact form, it is generally better to avoid an area of less than 5 or more than 45 acres. The purpose of the aforestated limits is to create fractional lots of dimensions that will facilitate all entries being made in a form that is optional with the entryman; an adherence to this practice will greatly reduce the necessity for the construction of supplemental plats now frequently demanded for no other purpose. Extreme lengths or narrow widths should be avoided; the longer direction should extend back from a meander line or claim boundary rather than along the same. It is inconsistent that a fractional lot lie partly in two sections, and it is generally better, when consistent with other rules, to avoid fractional lots extending from one into another fractional quarter section.

**199.** To secure a uniform system for numbering lots of fractional sections, including those above specified, imagine the section divided by parallel latitudinal lines into tiers, numbered from north to south; then, beginning with the eastern lot of the north tier, call it No. 1, and continue the numbering west through the tier, then east in the second, west in the third, east in the fourth tier, etc., until all fractional lots have been numbered. A lot extending north and south through two, or part of two tiers, will be numbered in the tier containing its greater area. In case any tier is without numbered lots, the numbering will be continued in the next tier to the south. This method of numbering will apply to any part of a section. A section that has been partly surveyed at different times should have no duplication of lot numbers.

**200.** When, by reason of irregular surveys or from other causes, the length of a township from south to north exceeds the regular length of 480 chains, or the width from east to west exceeds 480 chains, to such an extent as to require two or more tiers of lots along the north boundary, or two or more ranges of lots along the west boundary, as the case may be, the entire north or west portions of said sections beyond the regular legal subdivisions usually provided in these sections, will be suitably lotted, and to each lot will be assigned a proper number. Certain exceptions to this rule will be found in Chapter VII, in the instance of townships which possess abnormal dimensions in one or both directions.

**201.** If the first meridional section line of a township has been established as a sectional guide meridian, or the first latitudinal section line has been established as a sectional correction line, fractional lots will result along the east or south boundary of the township, as existing conditions may necessitate. Thus, where either the east or south boundaries of a township are defective in alinement (and not subject to rectification before subdividing) the sections bordering such defective boundaries will be subdivided by protraction in accordance with rules similar to those which operate in regard to sections bordering the north and west boundaries of a normal township. Other examples of subdivision of sections will be found under the general subject of "fragmentary subdivision."

SUBDIVISION BY SURVEY.

**202.** The rules for subdivision of sections by actual survey in the field are based upon the laws governing the survey of the public lands. When cases arise which are not covered by these rules, and the advice of the General Land Office in the matter is desired, the letter of inquiry should, in every instance, contain a description of the particular tract or corner, with reference to township, range and section of the public surveys, to enable the office to consult the record; also a diagram showing conditions found, giving distances in chains and links and not in feet.

**203.** Preliminary to subdivision it is essential to know the actual boundaries of the section, as it can not be subdivided legally until the section corners and quarter-section corners have either been found or restored by proper methods, and the resulting courses and distances determined by survey. The practice of entering a section to survey a tract from only one or two corners, and those perhaps unreliable, is unlawful.

**204.** The order of procedure is: First, identify or reestablish the boundary corners; next, fix the lines of quarter sections; then, form smaller tracts by equitable and proportionate division, according to the following rules:

**205.** *Subdivision of sections into quarter sections.*—Under the provisions of the act of Congress approved February 11, 1805, the course to be pursued in the subdivision of sections into quarter sections is to run straight lines from the established official quarter-section corners to the opposite corresponding corners. The point of intersection of the lines thus run will be the corner common to the several quarter sections, or, in other words, the legal center of the section.

Upon the lines closing on the north and west boundaries of a regular township the quarter-section corners are established by the United States surveyors at 40 chains to the north or west of the last interior section corners, and the excess or deficiency in the measurement is thrown into the half mile next to the township or range line, as the case may be.

Where there are double sets of section corners on township and range lines the quarter-section corners for the sections south of the township lines and east of the range lines have not always been established in the field by the United States surveyors, but in subdividing such sections said quarter-section corners should be so placed as to suit the calculations of the areas of the quarter sections adjoining the township boundaries as expressed upon the official plat, adopting proportionate measurements where the new measurements of the north or west boundaries of the section differ from the original measurements.

**206.** *Subdivision of fractional sections.*—The law provides that where opposite corresponding quarter-section corners have not been or can not be fixed, the subdivision-of-section lines should be ascertained by running from the established corners north, south, east or west lines, as the case may be, to the water course, reservation line, or other boundary of such fractional section, as represented upon the official plat. In this the law presumes the section lines surveyed and marked in the field by the United States surveyors to be due north and south or east and west lines, but this is not usually the case. Hence, in order to carry out the spirit of the law, it will be necessary in running the subdivisional lines through fractional sections to adopt mean courses, where the section lines are not due lines, or to run the subdivision-of-section lines parallel to the east, south, west or north boundary of the section, as conditions may require, where there is no opposite section line. (See sec. 197.)

**207.** *Subdivision of quarter sections into quarter-quarter sections.*—Preliminary to the subdivision of quarter sections, the quarter-quarter- or sixteenth-section corners will be established at points midway between the section and quarter-section corners, and between the quarter-section corners and the center of the section, except on the last half mile of the lines closing on irregular boundaries, where they should be placed at 20 chains, proportionate measurement, counting from the regular quarter-section corner.

The quarter-quarter- or sixteenth-section corners having been established as directed above, the center lines of the quarter section will be run straight between opposite corresponding quarter-quarter- or sixteenth-section corners on the quarter-section boundaries. The intersection of the lines thus run will determine the legal center of a quarter section.

**208.** *Subdivision of fractional quarter sections.*—The subdivisional lines of fractional quarter sections will be run from properly established quarter-quarter- or sixteenth-section corners, with courses governed by the conditions represented upon the official plat, to the lake, water-course or reservation which renders such tracts fractional. (See sec. 197.)

Fig. 49.

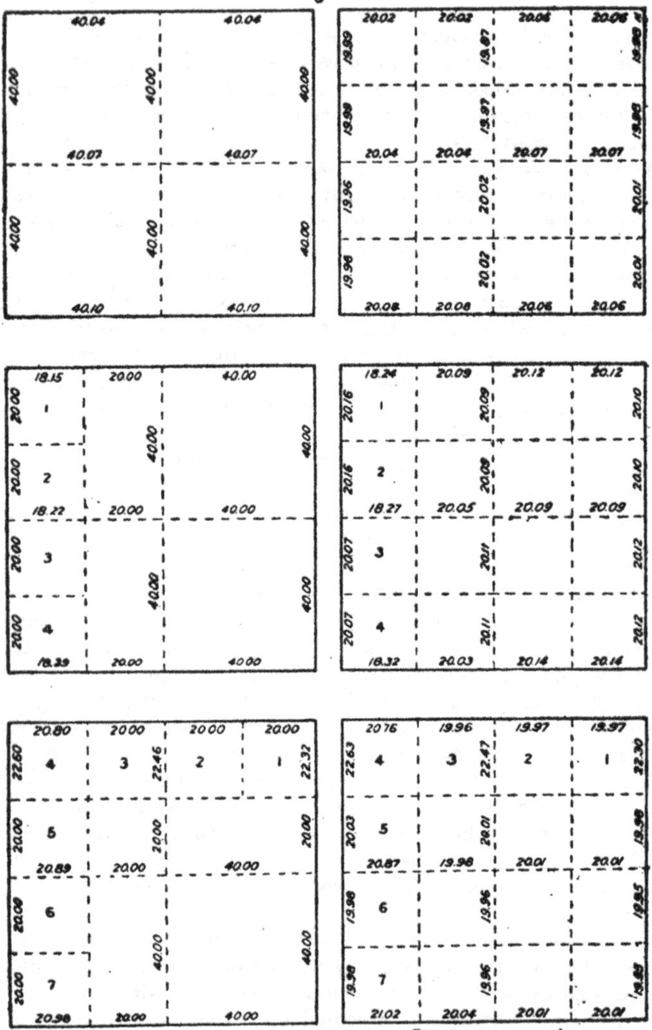

Official measurements. Remeasurements.

The above examples of subdivision by survey show the relation of the official measurements and calculated distances to the remeasurements, and indicate the proportional distribution of the differences.

**209.** By "proportionate measurement" is meant a measurement having the same ratio to that recorded in the original field notes as the length of the line by re-measurement bears to its length as given in the record. Reasonable discrepancies between former and new measurements may generally be expected. Errors may occur through many causes and should be as carefully avoided in re-measurements as in original surveys. Instead of the old practice of "adjusting the chain" to suit the former measure, the distance obtained by a precise method is compared with that of the record, and the shortage or surplus is computed by proportion, producing the same result in a more reliable manner. For example: The length of the line from the quarter-section corner on the west boundary of section 2 to the north line of the township, by the United States surveyor's measurement was reported as 43.40 chains, and by the county surveyor's measurement was found to be 42.90 chains; then the distance which the quarter-quarter- or sixteenth-section corner should be located north of the quarter-section corner would be determined by proportion as follows: As 43.40 chains, the official measurement of the whole distance, is to 42.90 chains, the county surveyor's measurement of the same distance, so is 20 chains, original measurement, to 19.77 chains by the county surveyor's measurement, showing that by proportionate measurement in this case the quarter-quarter- or sixteenth-section corner should be set at 19.77 chains north of the quarter-section corner, instead of 20 chains north of said corner, as represented on the official plat. In this manner the discrepancies between original and new measurements are equitably distributed.

**210.** By way of recapitulation it should be emphasized that when entrymen have acquired title to certain legal subdivisions they have become the owners of the identical ground area represented by the same subdivisions upon the official plat. It is a matter of expert or technical procedure to mark out the legal subdivisions called for in a patent, and entrymen are advised that a competent surveyor should be employed. The surveyor must necessarily identify the section boundaries and locate the legal center of the section in order to determine the boundaries of a quarter section. Then, if the boundaries of quarter-quarter sections, or fractional lots, are to be determined on the ground, the boundaries of the quarter section must be measured, and the sixteenth-section corners thereon should be fixed in accordance with the proportional distances represented upon the approved plat, thereupon the legal center of the quarter section may be duly located. Thus will be produced in the field the figure represented upon the plat, every part of the former in true proportion to the latter, where the elements of absolute distance and area have given away to corresponding proportional units as defined by fixed monuments established in the original survey.

### FRAGMENTARY SUBDIVISION OF TOWNSHIPS.

211. In the preceding articles covering the subject of subdivision of townships every assumption was based upon initiating the subdivisional survey upon regularly established exteriors, or, when necessary, a sectional guide meridian or a sectional correction line, or both, were to be established, upon which rested the control of the subdivision of the township. The subdivision of every full township may always be governed by the aforestated rules, but many other factors operate in determining the method and order of procedure to be adopted in the instance of fractional townships which have no linear south or east boundary, or in the case of continuing with the survey of partially subdivided townships, where one or more of the previously established section lines may be found to be defective in respect to the rectangular limit, or where partially surveyed sections, or sections containing outlying areas protracted as surveyed, are to be completed. The surveyor can not hope to master the subject of fragmentary subdivision of townships until he has become thoroughly familiar with every question relating to the subdivision of sections, nor is it possible to give in the Manual an example of every intricate problem which may be encountered in the field; thus the following discussion deals primarily with the principles, which must be considered in the field, operating to control the surveyor's method and order of procedure. It is possible, however, that cases may arise so complex in their character as to produce a feeling of doubt relative to the proper solution of the problem; in which case the surveyor will at once communicate with the proper supervising officer, submitting information, by letter and diagram, of the exact condition as found by him, and the necessary instructions will be forwarded as soon as practicable.

### FRACTIONAL TOWNSHIPS.

212. Where by reason of the presence of a large meanderable body of water, impassable objects, a State or reservation or grant boundary, or for other similar reasons a township is made fractional

Fig. 50.

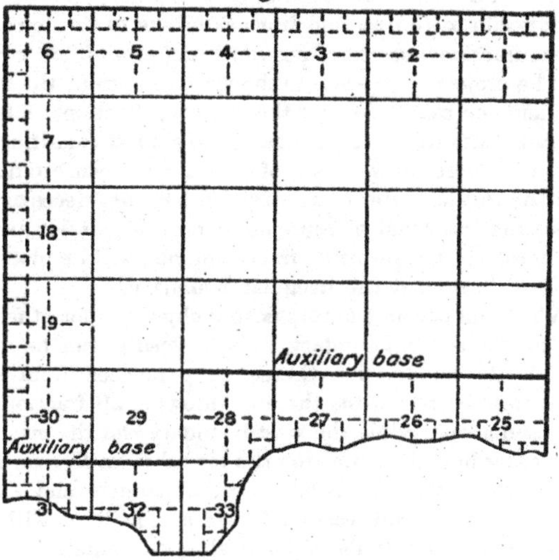
6
5
4
3
2
7
18
19
Auxiliary base
30
29
28
27
26
25
Auxiliary base
31
32
33

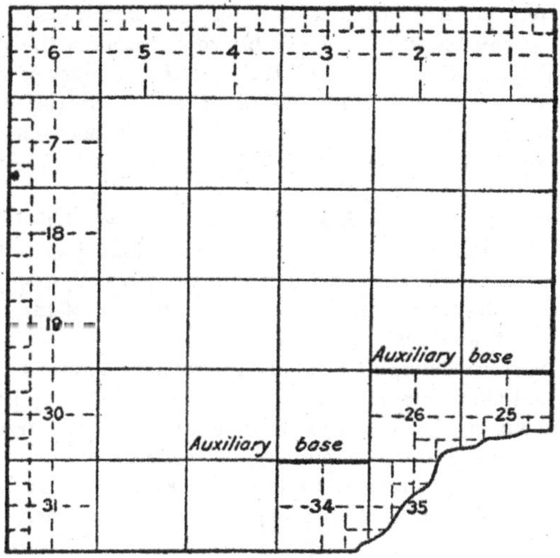
6
5
4
3
2
7
18
19
Auxiliary base
30
26
25
Auxiliary base
31
34
35

Fig. 51.

and is without a full linear south or east boundary, and it has been found advisable to run section lines as offsets to the township exteriors, the fractional section lines south and east of said controlling lines will be projected opposite to the usual direction; the fractional measurements on said lines and the resulting fractional lots will be placed against the irregular boundary. If similar conditions obtain throughout the north or west part of a fractional township no departure from the regular order of subdivision becomes necessary; in all such cases the fractional measurements on the exterior and subdivisional lines, and the resulting fractional lots, will be placed to the north and west against the irregular boundary.

**213.** Where on account of impassable objects or for other reasons no part of the south boundary of a township can be regularly established, the subdivision thereof may proceed from north to south and from east to west, thereby throwing all fractional measurements and areas against the west boundary and the meanderable stream or other boundary limiting the township on the south; if the east boundary is without regular section corners and the north boundary has been run eastwardly as a true line, with section corners at regular intervals of 80 chains, the subdivision of the township may be made from west to east, in which case the fractional measurements and areas will be thrown against the irregular east boundary; on the other hand, if the north boundary of section 6 is fractional, a sectional guide meridian will be initiated at the easternmost regular section corner on the north boundary of the township, which will be projected to the south to take the place of a governing east boundary, thus the subdivisional survey would be projected from north to south and from east to west, with fractional measurements, and resulting fractional lots, on the east, south and west boundaries of the township. The accompanying diagrams are illustrative of the principles which operate to control the subdivision of partial townships.

**214.** A very considerable class of surveys now coming before the General Land Office embraces the continuation of the subdivisional survey of townships previously subdivided in part only, frequently including the completion of partially surveyed sections or of sections containing outlying areas protracted as surveyed. If defective conditions are encountered in the previously established surveys, the problems concerning the procedure to be adopted multiply rapidly and require the greatest skill on the part of the surveyor.

Fig. 52.

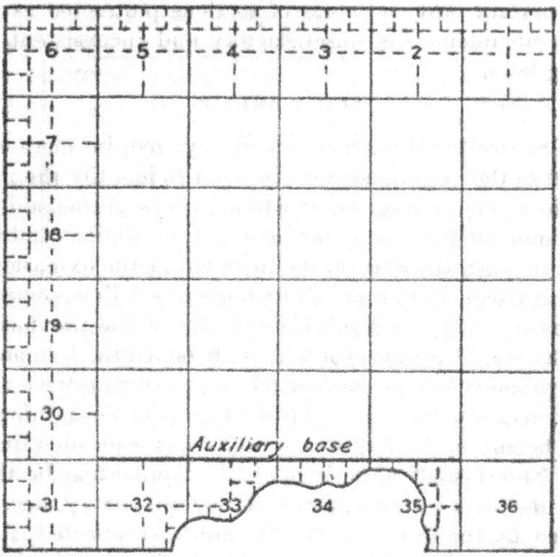


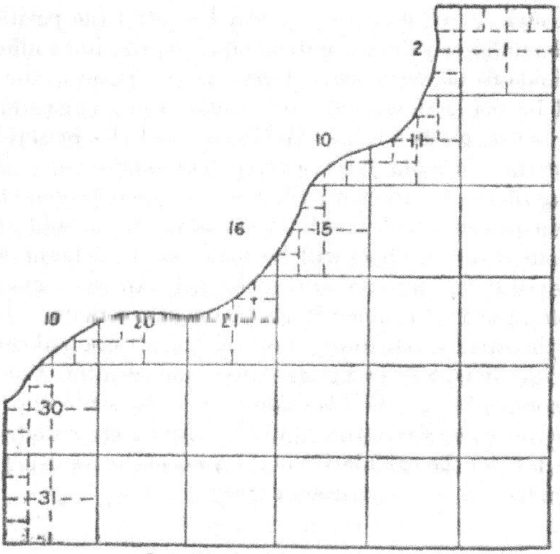


Subdivide Regularly.

Fig. 53.

In the construction of new township plats the former practice of showing certain outlying areas of sections protracted as surveyed has been abandoned as unsatisfactory and inconsistent with the surveying laws.

### RETRACEMENTS.

**215.** Practically all fragmentary surveys require more or less retracement of the original surveys in order to identify the initial and closing lines; such retracements will always be accompanied by the restoration of all lost corners adjacent to the sections embracing, in whole or in part, the areas to be included in the extension survey, in-so-far as the section or subdivision-of-section lines controlling the new areas may depend upon the position of the previously established corners. The surveyor will often be required, in order to determine properly the position of a lost corner, to retrace additional lines which are not the boundaries of sections containing the new areas to be surveyed, but no reestablishments on such lines are required. The theoretical position of a lost corner may be at variance with an unofficial corner established by local survey, accepted and recognized by the owners of the private lands affected; thus much trouble between landowners is avoided if the reestablishments are confined strictly to those corners which control the position of the section boundaries or the subdivision-of-section lines affecting the public lands to be surveyed. A general exception to the foregoing rule will be made in the case of identified original corners which are adopted as a basis from which to control the reestablishments bordering the public land sections; such original corners, if not in a good state of preservation, will be reconstructed in first-class order, a complete record of which will be embodied in the field notes. All restorations of lost corners will be made in strict accordance with the provisions of Chapter V of the Manual. In the instance of defective conditions contained in the previously established lines, exceeding the rectangular limit, even though all original corners may be fully identified and in a good state of preservation, the necessary retracements of the section boundaries will be made in order to determine the factors entering into the closing error and to furnish suitable data for the calculation of the areas of the resulting fractional lots embraced in the extension survey.

Fig. 54.

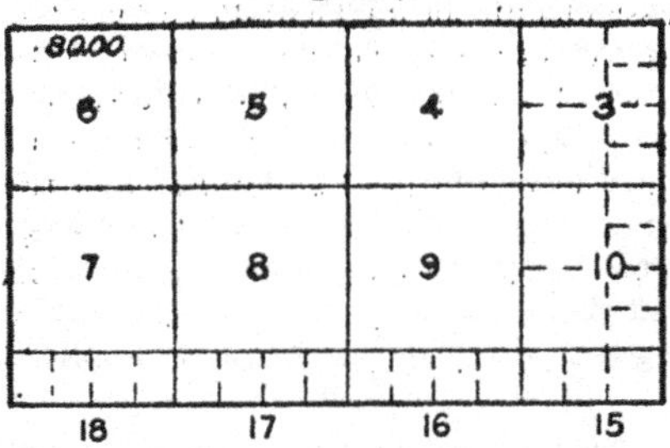


Subdivide from north to south, and from west to east.

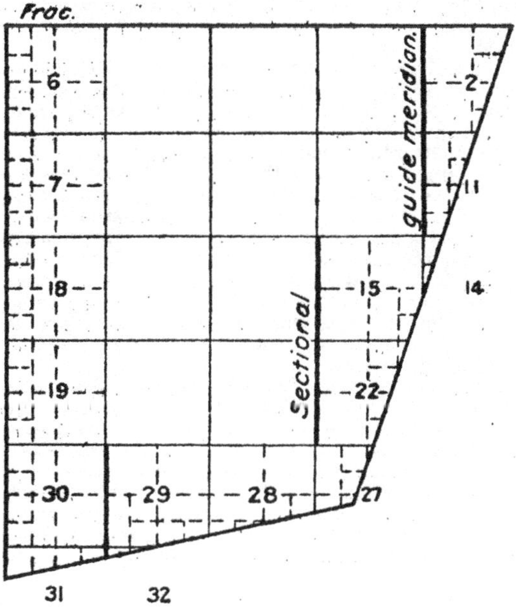


Subdivide from north to south, and from east to west.

## COMPLETION OF PARTIALLY SURVEYED SECTIONS.

**216.** Many assignments for fragmentary surveys require the completion of the survey of portions of boundaries of sections heretofore unsurveyed, in which sections are contained areas fixed in position by less than the regular complement of corners usually established for the identification of the legal subdivisions of the section. In the completion of such partially surveyed sections, the surveyor will be expected to give full consideration to the manner of protecting acquired rights based upon the former approved plats.

The following ten principles are distinctly applicable to the subject:

1st. The legal procedure governing the subdivision of any normal section into quarter sections is based broadly on the principle that the partition lines may be definitely fixed by four opposite quarter-section corners established on its boundaries; the intersection of the true center lines thus controlled is the legal point for the interior quarter-section corner of a section.

2d. The legal procedure governing the subdivision of regular quarter sections into quarter-quarter sections is based broadly on the same principle of controlling lines projected between opposite sixteenth-section corners of the quarter section, the latter corners established at mid-points on the true lines bounding the quarter section; the intersection of the true center lines of the quarter section is the legal point for the interior sixteenth-section corner of such regular quarter section.

3d. The legal procedure governing the subdivision of sections containing fractional lots into their component regular quarter-quarter sections and fractional lots is based on the same principle with the simple modification that the sixteenth-section corners on the boundaries of such quarter sections are themselves established at distances conformable to the proportions shown on the official plat.

4th. The fact that the full complement of four section corners of the section and all of the four opposite quarter-section corners has not been established in an accepted survey does not impair the validity of any areas shown upon the approved plat, and the legal procedure to be adopted in the extension of the boundaries of such sections must be such as to fix, within reasonable limits, the remaining quarter-section corners in a position which will protect the integrity of the original areas by controlling center lines connecting the old and new quarter-section corners.

5th. In the rectangular system the section is recognized as the unit of subdivision, and in proceeding with the extension of fragmentary surveys first consideration must necessarily be

Fig. 56.

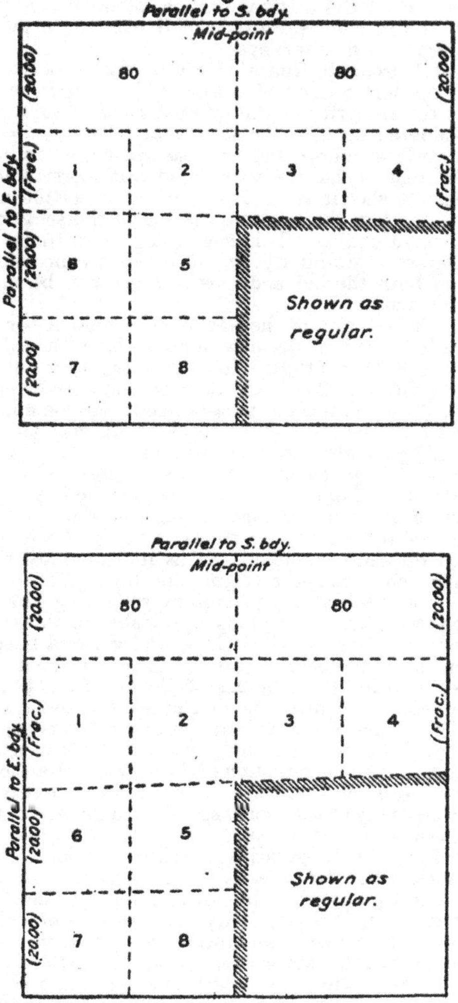


East boundary of section out of limits in measurement; southeast quarter protracted as surveyed; and section to be completed

given to the completion of the survey of fractional sections. No invasion of the original unit is tolerable if any portion of such unit has been surveyed, or if outlying areas have been shown protracted as surveyed.

6th. "Reasonable limits" for the fixation of the remaining quarter-section corners of a section in a position which will protect the integrity of the original areas of such section may be considered such as for alinement when not to exceed 21′ from a cardinal course, and for measurement when not to exceed 25 links from 40 chains where the opposite portion of the section boundary is shown as 40 chains, or in proportion as a limiting difference when the opposite portion of the section is more or less than 40 chains. This concession as to limits is made in the interest of simplicity, where by such concession rectangularity of both the old and new surveys may be maintained if so harmonized.

7th. The position of the new quarter-section corner which is to be established on the new opposite boundary of a fractional section will be controlled from one direction only if the old opposite distance has been made to count from one direction only, and the controlling measurement will be made to harmonize with the length of the opposite portion of the section, but if the old opposite distance has been made to count from two directions the position of the new quarter-section corner will be controlled from the two directions and the proportional lengths of the two portions of the new line will be made to harmonize with the proportional lengths of the two parts of the old opposite boundary, all as indicated by the distances and areas shown on the original approved plat.

8th. The underlying principles governing the rectangular surveying system are equally applicable to the completion of the survey of fractional sections, and given a condition in an original survey which in all its various elements is "within limits" within the meaning of the rectangular surveys, the simple plan of continuing in the same manner and order as would have been adopted in the original survey, if the same had not been discontinued, will accomplish usually in its simplest form the completion of the survey of fractional sections; this becomes the first duty of the surveyor before proceeding with the survey of additional sections, so that should irregularity be developed, no invasion of partially surveyed sections can result from the irregularities of other sections. It follows in principle, when irregularity is developed, that the surveyor will be best prepared to determine the proper method of survey adapted to procure simplicity of correction of existing irregularities and an early resumption of regularity, when he is in possession of full data concerning the conditions of all the old lines limiting the fragmentary surveys and upon which the new lines are to be initiated or closed, his knowledge being based upon the results of actual retracement of such irregular

Fig. 57.

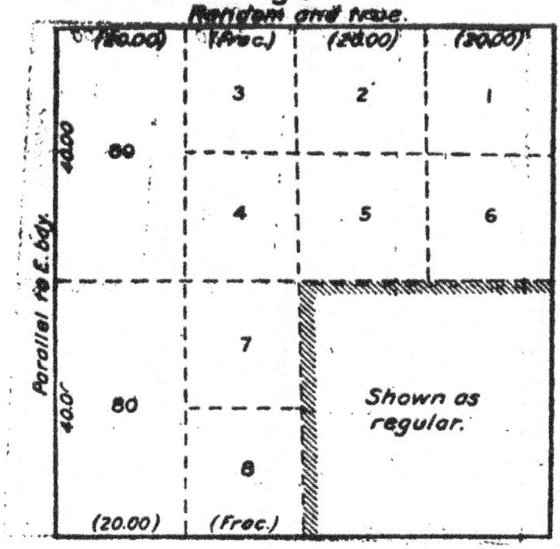


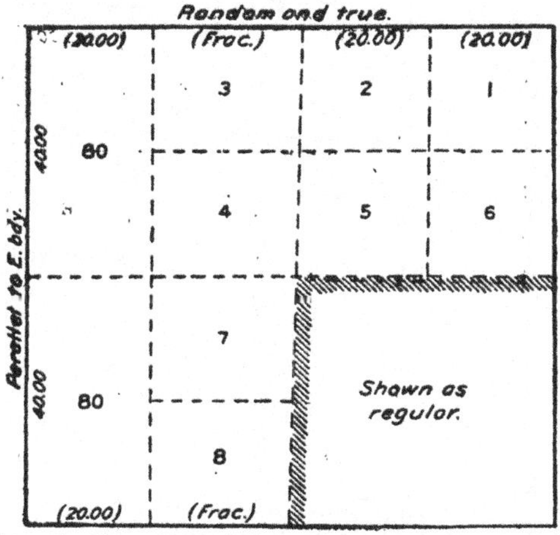


South boundary of section out of limits in measurement; southeast quarter protracted as surveyed; and section to be completed.

Fig. 58.

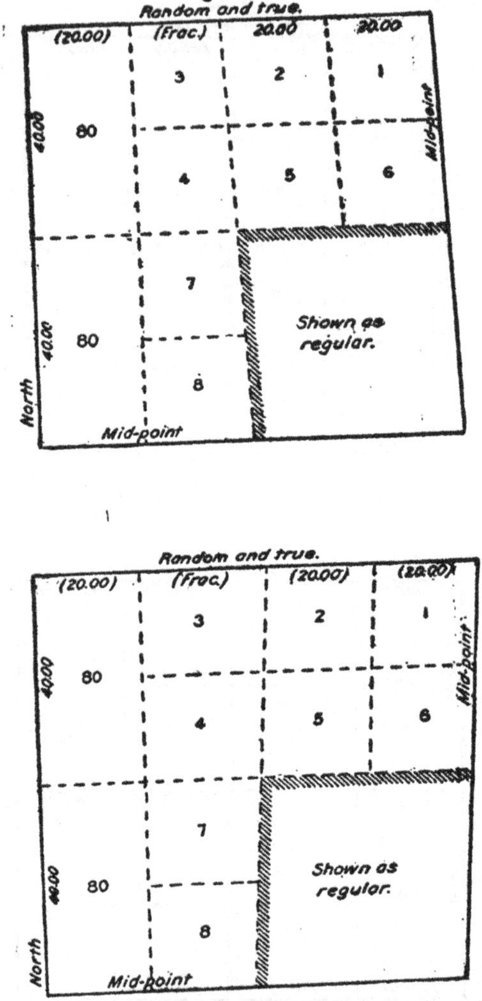


East boundary of section out of limits in alinement; southeast quarter protracted as surveyed; and section to be completed.

old lines. It must be granted that a skillful exercise of judgment by the surveyor based upon his knowledge of the facts is far more desirable than to restrict him to the application of empirical rules devised to cover possible, but innumerable combinations of irregularity.

9th. The completion of the survey of the partially surveyed sections will be made as nearly as possible in accordance with the regular rules for subdividing when the original lines are found to be within limits, otherwise, such sections will be completed by surveying all lines in such a manner that each and every section (excepting in cases of unavoidable hiatus or overlap) shall have four regular boundaries without offsets, with four governing section corners and four controlling quarter-section corners in such positions as to maintain the integrity of the fractional areas already shown upon the original plat. The subdivision thereof may then be made by connecting the opposite quarter-section corners in the regular manner with resulting locations agreeable to the legal subdivisions shown upon the original plat. If an hiatus or overlap is unavoidable, the position of the new quarter-section corner or corners will be carefully determined for latitude on a meridional line or for departure on a latitudinal line on the same plan as would have resulted in the regular survey of a new boundary extending in full from the one or two directions which control the position of the new quarter-section corner or corners.

10th. Adjoining sections must be considered separately when placing the new quarter-section corners, and the new corner need not be common to the four quarters of the two adjoining sections unless the theoretical position for each section falls within 25 links of a common point in which case the difference may be adjusted in such a manner as to secure maximum regularity.

**217.** Let it be assumed that adjacent to two established section lines, the *meridional* line of which is out of limits in measurement, an outlying regular quarter section has been protracted as surveyed; then to complete the section the new section lines will be extended from the previously established section corners, parallel to the opposite established boundaries, or mean course thereof, to a mutual intersection. The quarter-section corner on the new latitudinal section line would be established regularly at the mean point, and would ordinarily be marked to control the subdivision of two sections. On the new meridional boundary one or two quarter-section corners may be required; one marked to control the subdivision of the section under consideration will be established at 40 chains from the original section corner; the same quarter-section corner would be marked to control the subdivision of the adjoining section if the fractional

Fig. 59.

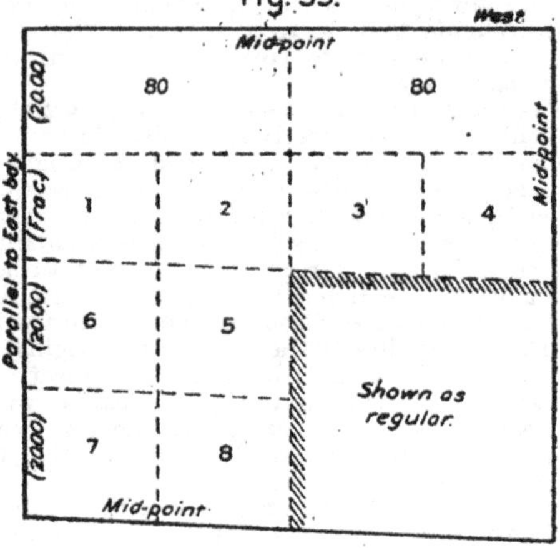


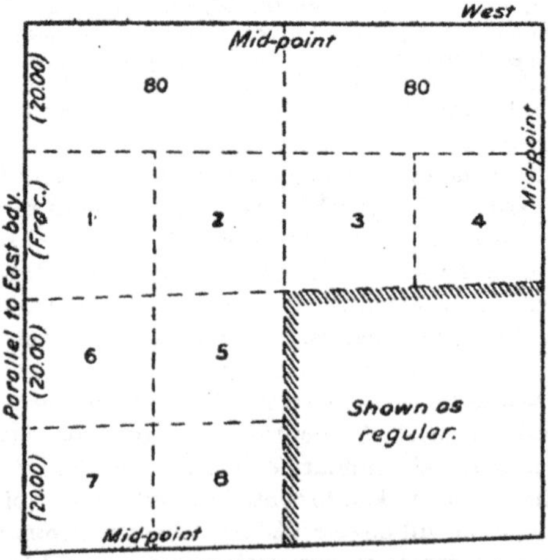


South boundary of section out of limits in alinement; southeast quarter protracted as surveyed; and section to be completed.

measurement is to be thrown in the same direction in the two sections, otherwise an additional quarter-section corner marked to control the subdivision of the adjoining section would ordinarily be placed at 40 chains from the new section corner. Again, let the same condition be assumed with the exception that the *latitudinal* section line instead of the meridional line is found to be defective in measurement. Then, to complete the section, the new meridional line would be surveyed as in regular subdivision, parallel to the opposite meridional line, or mean course thereof, ordinarily with quarter-section and section corners of maximum control at 40 and 80 chains, respectively. The new latitudinal section line would then be established on a true line between the section corners, and one or two quarter-section corners will be established as required; one marked to control the subdivision of the section under consideration will be established at 40 chains from the original section corner; the same quarter-section corner would be marked to control the subdivision of the adjoining section if the fractional measurement is to be thrown in the same direction in both sections, otherwise an additional quarter-section corner marked to control the subdivision of the adjoining section would ordinarily be placed at 40 chains from the new section corner.

**218.** Let another assumption be made that adjacent to two established section lines, the *meridional* line of which is out of limits in alinement, an outlying regular quarter section has been protracted as surveyed; then to complete the section, the new meridional line will be projected as a sectional guide meridian, in accordance with the usual rules, ordinarily with quarter-section and section corners of maximum control at 40 and 80 chains, respectively. The new latitudinal section line would then be established on a true line between the section corners, with one or two quarter-section corners as required; one marked to control the subdivision of the section under consideration will be required at 40 chains from the original section corner; the same quarter-section corner would be marked to control the subdivision of the adjoining section if the fractional measurement is to be thrown in the same direction in both sections; otherwise an additional quarter-section corner marked to control the subdivision of the adjoining section will ordinarily be established at 40 chains from the new section corner. On the other hand, if the same conditions be assumed with the exception that the original *latitudinal* section line instead of the meridional line is found to be

Fig. 60.

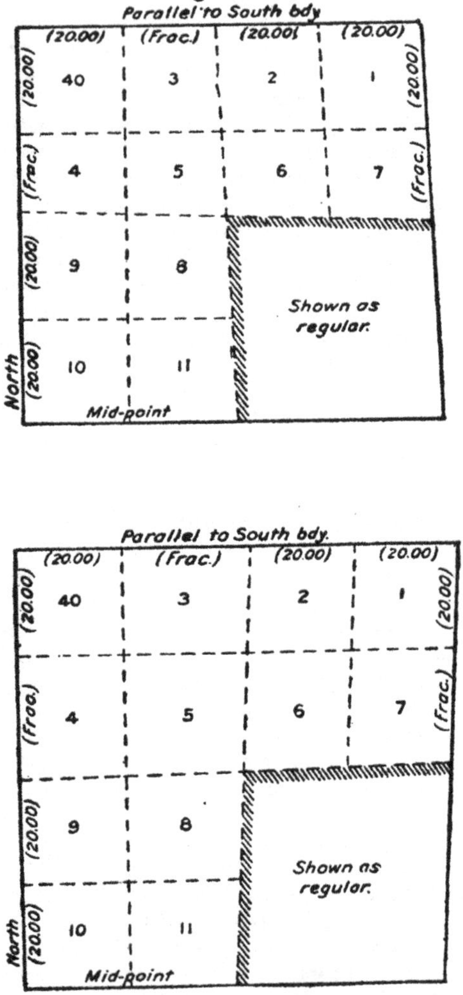


East boundary of section out of limits in alinement and measurement; southeast quarter protracted as surveyed; and section to be completed.

Fig. 61.

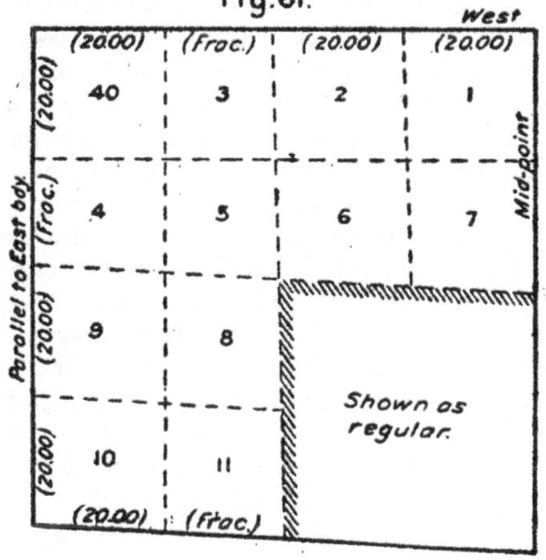


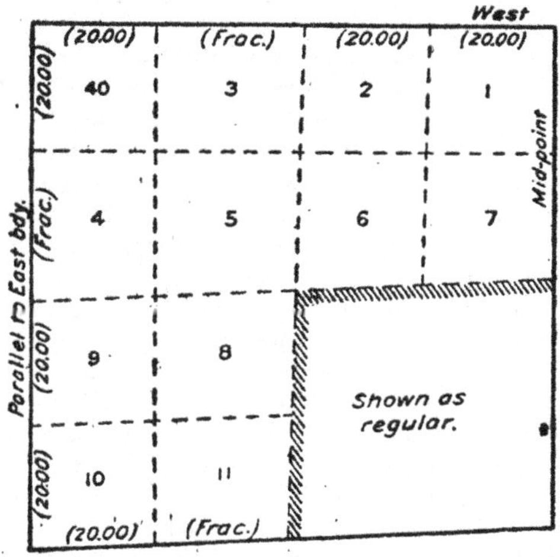


South boundary of section out of limits in alinement and measurement; southeast quarter protracted as surveyed; and section to be completed.

defective in alinement, then the new latitudinal section line will have to be established as a sectional correction line, exactly in accordance with the rules already given for running such lines, ordinarily with section corner of maximum control at its intersection with the new meridional section line, and quarter-section corner of maximum control at mid-point. On the new meridional section line one or two quarter-section corners may be required; one marked to control the subdivision of the section under consideration will be established at 40 chains from the original section corner; the same quarter-section corner may be marked to control the subdivision of the adjoining section if the fractional measurement is to be placed in the same direction in the two sections, but if the fractional measurement is to be thrown in the opposite direction in the adjoining section an additional quarter-section corner marked to control the subdivision of that section would ordinarily be required at 40 chains from the new section corner.

**219.** Many cases will arise in the field involving combinations of two or more of the above simple examples, in which instance the surveyor is advised to prepare a diagram illustrating the conditions found in the original survey, whereupon the new section lines may be shown with alinement in accordance with the usual rules for subdividing townships, noting that the new section lines are to be initiated at the previously established original section corners, and that the *length* of the meridional boundary will depend both upon the regularity of the length of the opposite original meridional section line and upon the alinement of the previously established latitudinal section line; thereupon the surveyor may at once show upon his diagram the position of the necessary quarter-section corners on the new section lines, all in conformity with the simple rules already stated.

**220.** Other instances will be found where half sections are shown upon the original approved plat protracted as surveyed, in some cases where only the opposite section line has not been established and in other cases where parts of the adjacent as well as the opposite section lines have not been established. In case only one section line remains to be established, it will be located upon the true line connecting the original section corners, regardless of bearing; the new opposite quarter-section corner marked to control the subdivision of the stated section will be placed at mid-point, regardless of the length of the new section line; the position of the quarter-

Fig. 62.

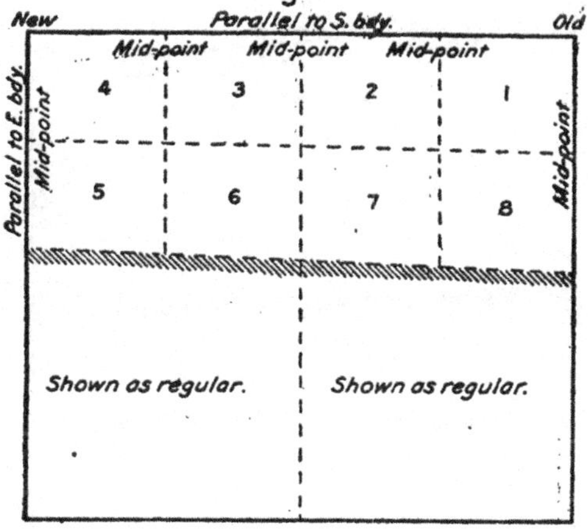


Old bdrs. defective in measurement.

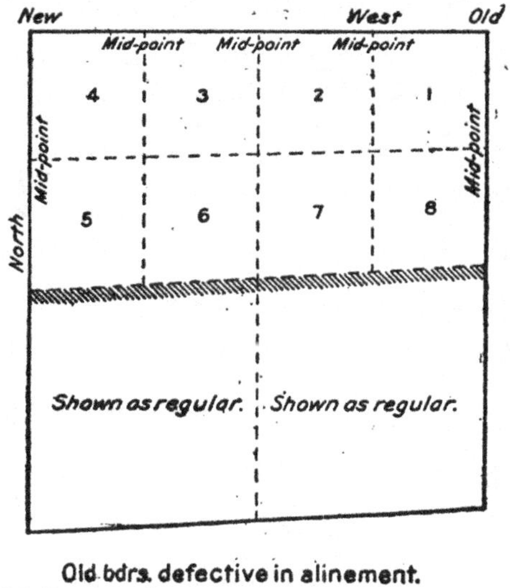


Old bdrs. defective in alinement.

South half protracted as surveyed, and section to be completed.

55465°—19——14

Fig. 63.

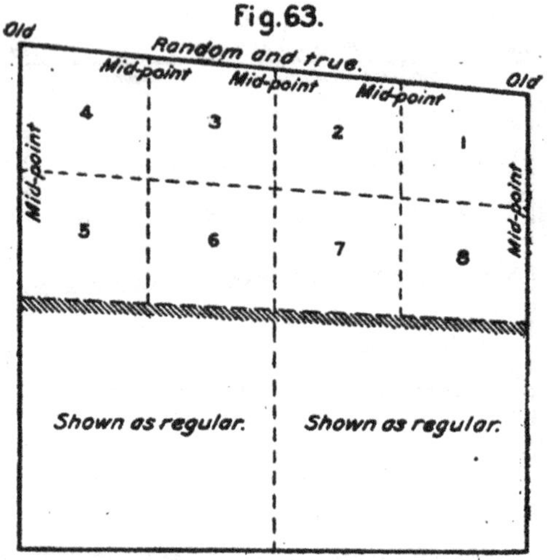


Old bdrs. defective in measurement.

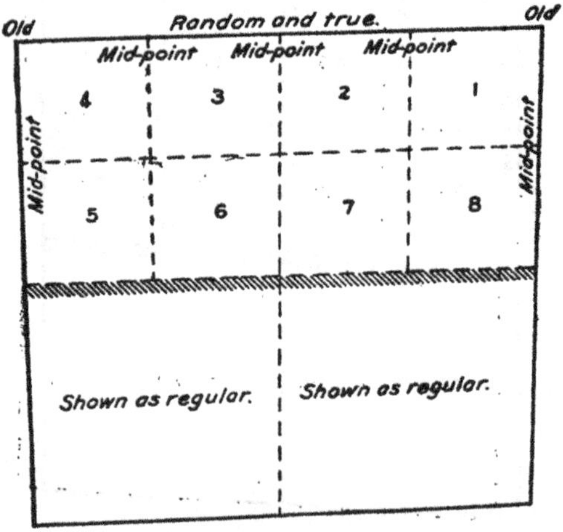


Old bdrs. defective in alinement.

South half protracted as surveyed, and section to be completed.

Fig. 64.

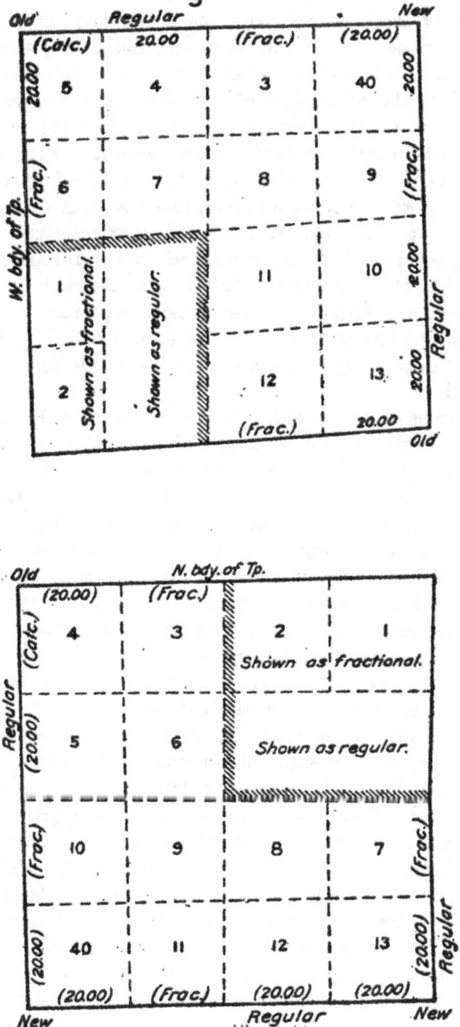


Old surveys irregular; protracted areas shown as fractional; and section to be completed.

section corner marked to control the subdivision of the adjoining section will depend upon the plan of subdividing the remaining public land. Partially surveyed section lines will be completed by extension, the alinement of the same being governed by the usual rules for regular subdivision; the latitudinal or meridional position of the remaining section line (opposite to the half section protracted as surveyed) will usually be controlled by the position of the nearest original section corner, and the alinement of the same will depend upon the usual rules for regular subdivision; the new opposite quarter-section corner marked to control the subdivision of the section containing such half section protracted as surveyed will be placed at mid-point in every case; the position of the quarter-section corner marked to control the subdivision of the adjoining section will depend upon the manner of subdividing the remaining public land.

**221.** Various other examples will be found where fractional areas, as along the north or west boundary of a township, are shown upon the original approved plat protracted as surveyed. In all such instances the same rules, heretofore stated, may be applied, with the single exception that a calculation must be made, based upon the areas shown upon the original plat, of the theoretical lengths of all lines not established in the original survey. Such calculated distances will then control instead of the usual regular lengths of section lines as heretofore assumed; also, if such calculated distances count from two directions, and irregularities are developed, the calculations must again be resolved into proportional distances to agree with actual measurements between the controlling points.

**222.** On the accompanying diagrams are shown various exaggerated examples of the manner of completing the survey of irregular sections containing outlying areas protracted as surveyed, showing the application of the means necessary for the protection of the integrity of such areas. It is recognized that the general principles above set forth will not always permit the complete establishment and appropriate marking of all corners at the first determination of their locations, by reason of the fact that only the bringing up of the new surveys to be closed upon the completed units will develop the appropriate markings of the finished corner, but this need not impair the surveyor's confidence in his knowledge of necessary procedure in the initiatory work, to be recognized and applied appropriately when the new surveys are brought up to their closings.

**223.** A distinctly different class of partially surveyed sections is found along erroneous meander lines shown upon approved plats of fractional townships. Such sections are never subject to completion except as definitely authorized in the written special instructions furnished to the surveyor, as the approved plat must be held to represent correctly a true meanderable body of water until proven otherwise to the satisfaction of the Department of the Interior, as intimated in Chapter I. Numerous instances are on record, however, where the evidence submitted to the Department is conclusive that surveyors have erroneously classified overflowed lands as meanderable, or where the recorded meander line does not and never did conform to the mean high-water elevation of an actual meanderable body of water, thus erroneously omitting considerable areas of land. The questions of title to such areas are extremely intricate, and it is the practice of the General Land Office not to allow any extension of such original surveys until the procedure has been definitely authorized by the Secretary of the Interior. The surveying problems arise only when the extension of the original survey beyond the meander line shown upon the approved plat has been duly authorized.

The reestablishment of the original meander line with a suitable monument at each angle point is a usual accompaniment of the above class of surveys, the purpose being to segregate definitely the previously surveyed areas from the unsurveyed public lands; it is more appropriate to consider the surveying questions thus involved along with other problems relating to the reestablishment of broken boundaries, where the subject will be found in sec. 380, Chapter V. The next step in the field is to complete the partially surveyed sections and the procedure in practically every instance will be controlled by the rules already outlined in respect to the completion of the survey of sections containing outlying areas protracted as surveyed; it seems unnecessary to repeat the governing principles in such closely related cases.

### SUBDIVISION OF FRACTIONAL SECTIONS RESULTING FROM FRAGMENTARY SURVEYS.

**224.** The one best test of the fitness of a proposed method incident to the completion of partially surveyed sections will be found in platting the section for subdivision by protraction; thereupon the regular rules for subdivision of sections should be applicable. Thus

Fig. 65 (West half)

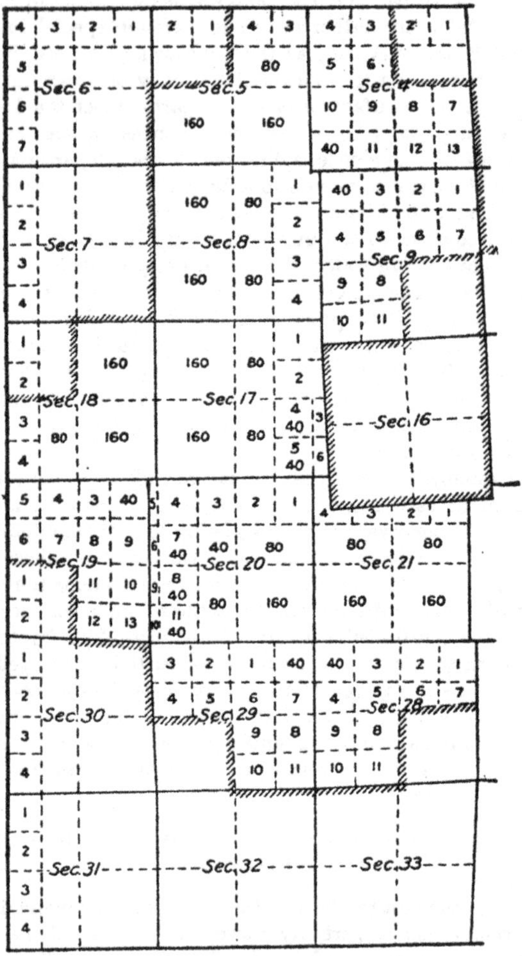


**Example showing the completion of partially surveyed sections, the subdivision of**
**resulting from**

Fig. 65 (East half)

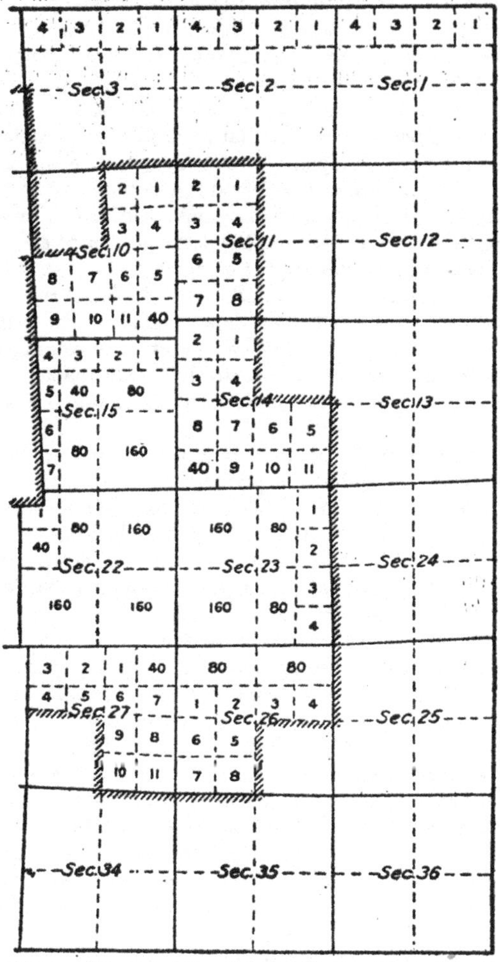


fractional sections, and the completion of the subdivisional lines of a partial township
fragmentary surveys.

the position of the new quarter-section corners, established to control the subdivision of a particular section in question, must be such as to permit the center lines from said points to the opposite original quarter-section corners to be connected in strict harmony with the conditions represented upon the original approved plat, disregarding the effect upon the subdivision of the newly surveyed public land. Likewise the lines connecting the sixteenth-section corners on the opposite boundaries of a quarter section must conform to the conditions represented upon the original plat. When the subdivision-of-section lines are thus platted the section may be considered satisfactory if the integrity of the original areas is in no way violated. When the subdivision-of-section lines are platted as suggested, the permanent conditions affecting the new areas may be considered, and should be harmonized with the following additional rules:

1st. The new areas should be complementary to the original areas by the extension of the subdivision-of-section lines as already protracted upon the original plat, except as poorly shaped lots, or lots of too great or too little area, would result in violation of the regular rules for subdivision of sections.

2d. The same meridional limit may be permitted, in the interest of regularity and simplicity of platting, as is ordinarily allowed in latitudinal section lines; i. e., a section may be considered regular whose boundary lines are all for alinement when not to exceed 21′ from a cardinal course, and for measurement when not to exceed 25 links from 40 chains between the section and quarter-section corners. Such regular sections may be subdivided into regular quarter sections and quarter-quarter sections as far as possible. A section having three regular boundary lines may be subdivided in accordance with the usual rules for subdividing sections along the north and west boundaries of a normal township. A section having two adjacent regular boundary lines may be subdivided similarly to the manner in which section 6 of a normal township is treated. All other sections should be treated as irregular, with subdivision-of-section lines protracted to mid-points on the boundaries of the quarter sections, except as a calculated proportional position for a sixteenth-section corner is made necessary by reason of conditions relating to the complementary area shown upon the original plat.

3d. All new fractional lots will be numbered beginning with the next higher number in the series of the same section already begun upon the previously approved plat, and proceeding in the usual order in which fractional lots are normally numbered. The new series may begin with No. 1 in case the fractional parts of the original area are not designated by lot number.

### COMPLETING THE SUBDIVISION OF A PARTIAL TOWNSHIP RESULTING FROM FRAGMENTARY SURVEYS.

**225.** After the partially surveyed sections have been fully completed the surveyor may proceed with the subdivision of the remaining portions of the township. Every condition represents a separate problem, and few specific rules would serve any purpose in guiding the surveyor to a definite procedure. If no irregularities are to be found in the previously established lines the new survey may proceed normally, but if defective conditions are encountered the irregularities are not to be extended into unsurveyed sections any farther than necessary to incorporate the resulting fractional measurements into suitable fractional lots adjoining the former surveys. Preference should be given to extending all surveys from south to north and from east to west, but if a better control is available by reversing the procedure in one or both directions, thus resulting in a simpler and better survey in respect to minimizing the number of extra corners as well as fractional lots, such reversal of procedure is fully warranted. The principle relating to controlling coördinate measurements in two directions at right angles, as along the south and east boundaries of a township, may be applied to the subdivisional lines best suited to control the new surveys to be executed; and, if the selected bases are defective in alinement, in whole or in part, the new section lines may serve the function of a sectional guide meridian or a sectional correction line as required. The corners from which the new surveys are to be initiated and controlled in latitude and departure will be termed corners of four sections, or of two sections as appropriate, and where the terminal lines can not be connected regularly with the previously established section corners by random and true line not exceeding 21′ from cardinal, a closing section corner will be established in full accord with the principle relating to the establishment of closing section corners on the north or west boundaries of a township where the latter lines are found to be defective in measurement. The fractional measurements of the closing section lines will be placed adjacent to the old surveys, and the distance from the closing section corner to the nearest original corner will

Fig. 66 (West half)

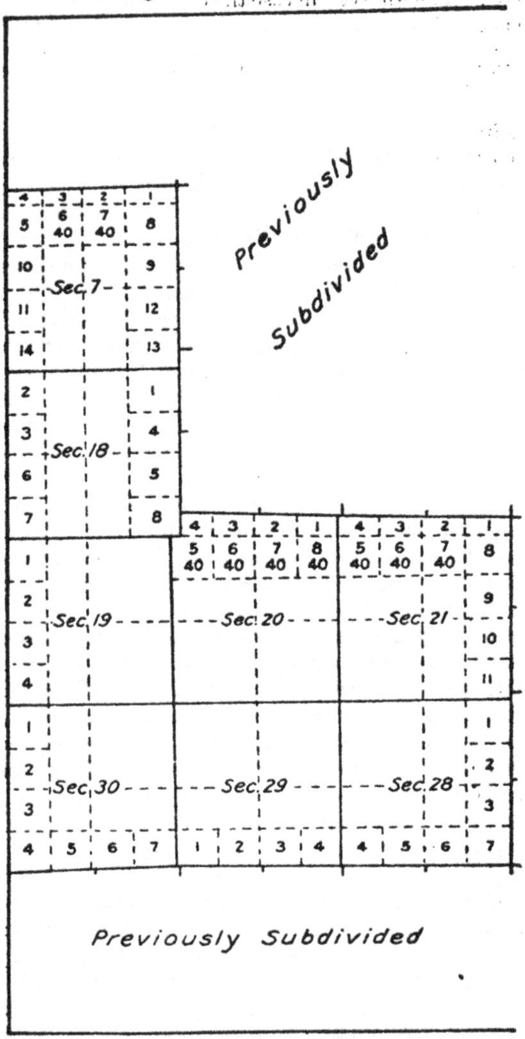


Example showing the completion of the subdivisional lines

Fig. 66 (East half)

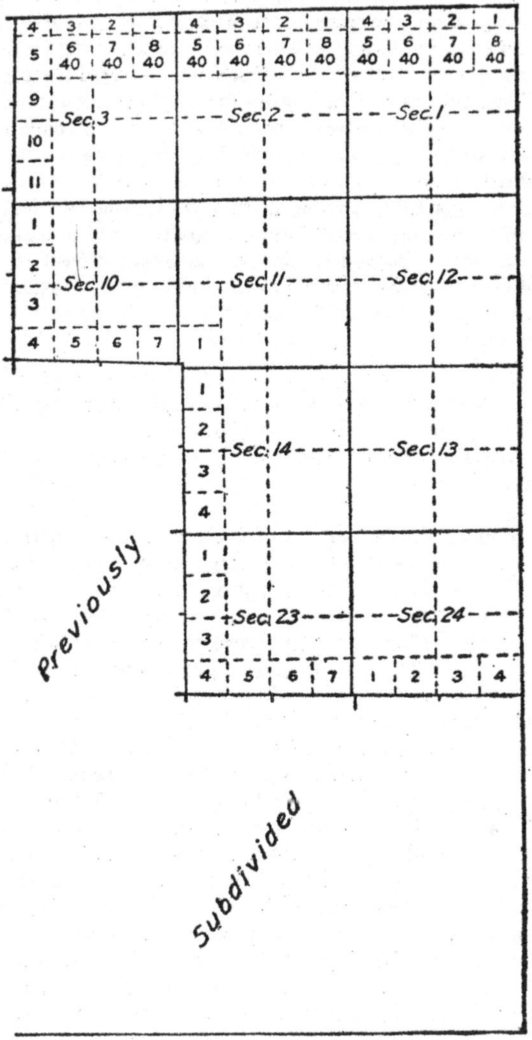


of a partial township resulting from fragmentary surveys.

be measured; the original lines forming the boundary of the lands to be surveyed will be retraced, as already provided, and the marks upon the original corners will be appropriately modified as necessary; new quarter-section corners marked to control the subdivision of the new sections will be established on the original lines at midpoints between the closing section corners, or at 40 chains from one direction, according to the manner in which a new section is to be subdivided.

There are generally two or more ways in which a fragmentary subdivision may be executed, but a careful study of a sketch plat representing existing conditions will generally reveal the superiority of one method over another, and objectionable results should be avoided as far as existing conditions relating to the original surveys will permit.

### MEANDERING.

**226.** All navigable bodies of water and other important rivers and lakes (as hereinafter described) are to be segregated from the public lands at mean high-water elevation. The traverse of the margin of a permanent natural body of water is termed a meander line.

The running of meander lines has always been authorized in the survey of public lands fronting on large streams and other bodies of water, but the mere fact that an irregular or sinuous line must be run, as in case of a reservation boundary, does not entitle it to be called a meander line except where it closely follows the bank of a stream or lake. The legal riparian rights connected with meander lines do not apply in case of other irregular lines, as the latter are strict boundaries.

Mean high-water mark has been defined in a State decision (47 Iowa, 370) in substance as follows: High-water mark in the Mississippi River is to be determined from the river bed; and that only is river bed which the river occupies long enough to wrest it from vegetation. In another case (14 Penn. St., 59) a bank is defined as the continuous margin where vegetation ceases, and the shore is the sandy space between it and low-water mark.

Numerous decisions in the United States Supreme Court and many of the State courts assert the principle that meander lines are not boundaries defining the area of ownership of tracts adjacent to waters. The general rule is well set forth (10 Iowa, 549) by saying that in a navigable stream, as the Des Moines River in Iowa, high-water mark is the boundary line. When by action of the water the river bed changes, high-water mark changes and ownership of adjoining land progresses with it.

Meander lines will not be established at the segregation line between upland and swamp or overflowed land, but at the ordinary high-water mark of the actual margin of the river or lake on which such swamp or overflowed lands border.

**227.** Practically all inland bodies of water pass through an annual cycle of changes from mean low water to flood stages, between the extremes of which will be found mean high water. In regions of broken topography, especially where bodies of water are bounded by sharply sloping lands, the horizontal distance between the margins of the various water elevations is comparatively slight, and the surveyor will not experience much difficulty in determining the horizontal position of mean high-water level with approximate accuracy; but in level regions, or in any locality where the meanderable bodies of water are bordered by relatively flat lands, the horizontal distance between the successive levels is relatively great. The surveyor will find the most reliable indication of mean high-water elevation in the evidence made by the water's action at its various stages, which will generally be found well marked in the soil, and in timbered localities a very certain indication of the locus of the various important water levels will be found in the belting of the native forest species.

Mean high-water elevation will be found at the margin of the area occupied by the water for the greater portion of each average year; at this level a definite escarpment in the soil will generally be traceable, at the top of which is the true position for the surveyor to run the meander line. A pronounced escarpment, the result of the action of storm and flood waters, will often be found above the principal water level, and separated from the latter by the storm or flood beach; another less evident escarpment will often be found at the average low-water level, especially of lakes, the lower escarpment being separated from the principal escarpment by the normal beach or shore. While these questions properly belong to the realm of geology, they should not be overlooked in the survey of a meander line.

Where native forest trees are found in abundance bordering bodies of water, those trees showing evidence of having grown under favorable site conditions will be found accurately belted along contour lines; thus a certain class of mixed varieties common to a particular region will be found only on the lands seldom if ever overflowed; another group of forest species will be found on the lands which are inundated only a small portion of the growing season each year, and indicate the area which should be included in the classification of the uplands; other varieties of native forest trees will be found only within the zone of swamp and overflowed lands. All timber growth normally ceases at the margin of permanent water.

**228.** At every point where either standard, township or section lines intersect the bank of a navigable stream, or any meanderable body of water, corners at such intersections will be established at the time of running these lines. Such monuments are called meander corners. In the survey of lands bordering on tide waters, meander corners may be temporarily set at the intersection of the surveyed lines with the margin of mean high tide, but no monument should be placed in a position exposed to the beating of waves and the action of ice in severe weather. In all such cases a witness corner on the line surveyed, at a secure point near the true point for the meander corner, will be established. The crossing distance between meander corners on the same line will be ascertained by triangulation or direct measurement, and the full particulars will be given in the field notes.

**229.** Inasmuch as it is not practicable in public-land surveys to meander in such a way as to follow and reproduce all the minute windings of the high-water line, the United States Supreme Court has given the principles governing the use and purpose of meandering shores in its decision in a noted case (R. R. Co. *v.* Schurmeir, 7 Wallace, 286–287) as follows:

"Meander lines are run in surveying fractional portions of the public lands bordering on navigable rivers, not as boundaries of the tract, but for the purpose of defining the sinuosities of the banks of the stream, and as the means of ascertaining the quantity of land in the fraction subject to sale, which is to be paid for by the purchaser. In preparing the official plat from the field notes, the meander line is represented as the border line of the stream, and shows to a demonstration that the water-course, and not the meander line as actually run on the land, is the boundary."

**230.** The surveyor will commence the meander line at one of the meander corners, follow the bank or shore line, and determine the true bearing and measure the exact length of each course, from the beginning to the next meander corner. All meander courses are to be taken or counted from the true meridian and will be determined with precision; "transit angles" showing only the amount of the deviation from the preceding course are not acceptable in field notes of meanders. For convenience the courses of meander lines should be adjusted to the exact quarter degree; meander lines are not strict boundaries and this method will give approximate agreement with the minute sinuosities of mean high-water elevation. Again, for convenience of platting and computation, the surveyor is required to adopt turning points at distances of whole chains, or multiples of ten links, with odd links only in the final course.

In cases where the surveyor finds it impossible to carry his meander line along mean high-water mark, his notes should state the distance therefrom and the obstacles which justify the deviation. A table of latitudes and departures of the meander courses should be computed before leaving the vicinity, and if misclosure is found, indicating error in measurement or in reading courses, the lines should be rerun.

All streams flowing into a river, lake or meanderable bayou will be noted, and the width at their mouths stated; also, the position, size and depth of springs, whether the water be pure or mineral; also, the heads and mouths of all bayous, all rapids and bars, will be noted, with intersections to the upper and lower ends of the latter, to establish their exact situation. The elevation of the banks of lakes and streams, the height of falls and cascades, and the length and fall of rapids, will be recorded in the field notes.

The field notes of meanders will show the corners from which the meanders commenced and upon which they closed, and will exhibit the meanders of each fractional section separately; following, and composing a part of such notes, will be given a description of the adjoining land, soil and timber, and the depth of inundation to which the bottom land is subject. The utmost care will be taken to pass no object of topography, or change therein, without giving a particular description thereof in its proper place in the notes of the meanders.

### RIVERS.

**231.** Proceeding downstream, the bank on the left hand is termed the left bank and that on the right hand the right bank. These terms will be universally used to distinguish the two banks of a river or stream. Navigable rivers and bayous, as well as all rivers not embraced in the class denominated "navigable," the right-angle width of which is 3 chains and upwards, will be meandered on both banks, at the ordinary mean high-water mark, by taking the general courses and distances of their sinuosities. Rivers not classed as navigable will not be meandered above the point where the average right-angle width is less than 3 chains, except that streams which are less than 3 chains wide and which are so deep, swift and dangerous as to be impassable may be meandered, where good agricultural lands along the banks require their separation into fractional lots for the benefit of settlers.

Shallow fresh-water streams, without any well-defined channel or permanent banks, will not be meandered. Tidewater streams, whether more or less than 3 chains wide, should be meandered at ordinary high-water mark, as far as tidewater extends.

### LAKES.

**232.** The meanders of all lakes of the area of 25 acres and upwards, will be commenced at a meander corner and continued, as above directed for navigable streams; from said corner, the courses and distances of the entire margin of the same, and the intersections with all meander corners established thereon, will be noted.

In the case of lakes which are found to be located entirely within the boundaries of a section, a quarter-section line, if one crosses the lake, will be run from one of the quarter-section corners, on a theoretical course to connect with the opposite quarter-section corner, to the margin of the lake, and the distance will be measured; then at the point thus determined a "special meander corner" will be established. If a meanderable lake is found to be located entirely within a quarter section, an "auxiliary meander corner" will be established at some suitable point on its margin, and a connecting line will be run from said monument to a regular corner on the section boundary. A connecting traverse line will be recorded, if one is run, but it will also be reduced to the equivalent direct connecting course and distance, all of which will be stated in the field notes, and the course and length of the direct connecting line will be be shown on the plat of the survey.

The meander line of a lake lying within the interior of a section will be initiated at the established special or auxiliary meander corner, as the case may be, and continued around the margin of the normal lake at its mean high-water level; to a closing at the point of beginning. All proceedings are to be fully entered in the field notes.

Artificial lakes and reservoirs are not to be segregated from the public lands, unless specially provided in the instructions, but the true position and extent of such bodies of water will be determined in the field and shown on the plat.

### ISLANDS.

233. In the progress of the regular surveys every island above the mean high-water elevation of any meanderable body of water, excepting only those islands which may have formed in navigable bodies of water after the date of the admission of a State into the Union, will be definitely located by triangulation or direct measurement or other suitable process, and will be meandered and shown upon the official plat.

In the survey of the mainland fronting on any non-navigable body of water, any island opposite thereto, above mean high-water elevation, is subject to survey. Also, even though the United States may have parted with its title to the adjoining mainland, an island in any meandered body of water, navigable or non-navigable, known or proven to have been in existence at the date of the admission of a State into the Union, and at the date of the survey of the mainland, if omitted from said original survey, remains public land of the United States, and as such the island is subject to survey.

The survey of islands not shown upon the original approved plats of subdivided townships is authorized by the Department only upon the receipt of formal application, and subject to the approval thereof. The proof of the time of the formation of such islands is often more or less difficult, and it is the practice of the Department to make a careful examination of the history of an island in relation to the question of its legal ownership before approving the application for its survey.

Any township boundary or section line which will intersect an island will be extended as nearly in accordance with the plan of regular surveys as conditions will permit, and the usual township, section, quarter-section and meander corners will be established on the island. If an island falls in two sections only, the line between those particular sections should be established in its proper theoretical position based upon suitable sights and calculations.

If an island falls entirely in one section, and is large enough to be subdivided (over 50 acres in area), a suitable sight or calculation will be made to locate on the margin of the island an intersection with the theoretical position of any suitable subdivision-of-section line, and at the point thus determined a "special meander corner" will be established. In the case of an island falling entirely in one section and found to be too small to be subdivided, an "auxiliary meander corner" will be established at any suitable point on its margin, which will be accurately connected with any regular corner on the mainland. The direct course and length of the connecting line will be given in the field notes, together with all sights, measurements, triangulations and traverse lines upon which the calculation may be based. The course and length of the direct connecting line will be shown on the plat.

The meander line of an island will be surveyed in harmony with principles and rules heretofore stated; all township and section lines crossing the island will be shown on the plat; and, if the island is large enough to be subdivided, the subdivision will be accomplished by the protraction of suitable subdivision-of-section lines in their correct theoretical position.

Agricultural upland within the limits of swamp and overflowed lands should be so classified and shown upon the plat accordingly, but such land will not be meandered as an island.

### LIMITS OF CLOSURE.

**234.** Under the general subjects of "township exteriors" and "subdivision of townships" certain definite limits were prescribed beyond which previously established surveys are classed as "defective," or in the case of new surveys corrective steps are required. Such limits constitute the standard of accuracy of the United States rectangular surveys, and, for convenience, have been variously referred to as the "rectangular limit," "limit for the control of new surveys," "limit relating to defective exteriors and section lines," "limits for subdivision," etc., each expression having been formed to suit the descriptive exigency of the text. A more general requirement known as the "limit of closure" will be applied as a test of the accuracy of the alinement and measurement of all classes of lines embraced in any closed figure incident to the public-land surveys, and corrective steps will be required wherever this test discloses an error beyond the allowable limit.

The "error of closure" of a survey may be defined, in general terms, as the ratio of the length of the line representing the equivalent of the errors in latitude and departure (as found by a table of latitudes and departures) to the length of the perimeter of the figure constituting the survey; but, with due regard for the controlling coördinate governing lines of a rectangular survey, pronounced accuracy in latitude will not be permitted to offset gross error in departure, or vice versa, and, in order to be consistent with this fundamental theory, a double test must be applied in place of the one expressed in general terms. The "limit of closure" fixed for the United States rectangular surveys may be expressed by the fraction $\frac{1}{452}$ provided that the limit of closure in neither latitude nor departure exceeds $\frac{1}{640}$, and where a survey qualifies under the latter limit the former is bound to be satisfied; thus an accumulative error of 12½ links per mile of perimeter, in either latitude or departure, will not be exceeded in an acceptable survey. The limit of closure as thus expressed may be applied to various specific conditions as heretofore stated.

The latitudes and departures of a normal section shall each close within 50 links; of a normal range or tier of sections, within 175 links; and of a normal township, within 300 links. The boundaries of each fractional section including irregular claim lines or meanders, or the meanders of an island or lake in the interior of a section, should close within a limit to be determined by the fraction $\frac{1}{640}$ when the error in either latitude or departure is considered separately; the same rule will be applied to all broken or irregular boundaries.

Surveyors are required to compute all doubtful closings while in the field in the immediate vicinity of a particular line, or series of lines, in question, and to accomplish all necessary corrective work before concluding a survey.

### MARKING LINES BETWEEN CORNERS.

**235.** The marking of a survey upon the ground in such a manner as to fix forever the position of the legal lines in relation to the earth's surface is the final step in the field work, and is accomplished in three ways, which, if well executed, will individually or collectively furnish the means of the identification of the survey at even remote future dates. Careful attention to these details is one of the most important phases of the surveyor's field work. (*a*) The regular corners of the public-land surveys are marked by fixed monuments of specified character as described in Chapter IV; (*b*) the relation of the officially surveyed lines to natural topographical features is recorded in much detail as hereinafter outlined, and again exemplified in the specimen field notes; and, (*c*) the locus of the legal lines, wherever living timber is encountered, is plainly marked upon the forest trees, which is accomplished by the process of "blazing" and by "hack" marks.

A "blaze" is an ax mark which is made upon a tree trunk at about breast height, in which a flat scar is left upon the tree surface. The bark and a very small amount of the live wood tissue are removed, leaving a smooth surface which forever brands the tree. The size of the blaze depends somewhat upon the size of the tree, but is never made larger than the surface of an ax blade; a blaze 5 or 6 inches in height and from 2 to 4 inches in width is ample to mark any tree.

A "hack" is also an ax mark which is made upon a tree trunk at about breast height, in which a horizontal notch is cut into the surface of the tree. The notch is made "V-shaped," and is cut through the bark and well into the wood. Two hacks are cut in order to distinguish those made in the survey from accidental marks resulting from other causes; a vertical section of the completed official hack mark resembles a "double-V" (≦) extending across a tree from 2 to 6 inches in length, depending upon the diameter of the tree. The "hack" and "blaze" marks are equally permanent, but so different in character that one mark should never be mistaken for the other.

The marking of trees along the surveyed lines was required by law as positively as the erection of monuments, by the act of 1796, which is still in force. All lines on which are to be established the legal corners will be marked after this method, viz: Those trees which may be intersected by the line will have two hacks or notches cut on each of the sides facing the line, without any other marks whatever. These are called sight trees or line trees. A sufficient number of other trees standing within 50 links of the line, on either side of it, will be blazed on two sides quartering toward the line, in order to render the line conspicuous, and readily to be traced in either direction, the blazes to be opposite each other coinciding in direction with the line where the trees stand very near it, and to approach nearer each other toward the line the farther the line passes from the blazed trees.

Due care will ever be taken to have the lines so well marked as to be readily followed, and to cut the blazes plainly enough to leave recognizable scars as long as the trees stand. This can be accomplished by blazing just through the bark into the live wood tissue. Where trees 2 inches or more in diameter occur along a line, the required blazes will not be omitted. Where trees have branches growing to the ground, the blazes will be omitted unless it is necessary to remove the branches to permit sighting.

Lines are also to be marked by cutting away enough of the undergrowth to facilitate correct sighting of instruments. Where lines cross deep wooded valleys, by sighting over the tops, the usual blazing of trees in the low ground when accessible will be performed, that settlers may find their proper limits of land and timber without special survey. The undergrowth will be especially well cut along all lines within distances of 5 chains of corner monuments and within 2 chains of arteries of travel, to enable other surveyors and settlers to locate the survey readily, but the cutting of the undergrowth may be omitted in deep untraveled ravines unless necessary for accurate sighting or measurement.

Line trees and blazing will be marked only with reference to the established true line, and where lines are run by the "random and true" line method, the marking of line trees and the blazing will be accomplished by returning over the line after all corrections or adjustments to the final line are definitely known. A sufficient number of temporary stakes should be set along a random line to render it generally unnecessary to rerun the true line instrumentally merely for the purpose of blazing the line through timber, as this can usually be accomplished by properly estimating the distance from the temporary stakes, but intersections with line trees will be made with precision, and distances thereto accurately measured.

## SUMMARY OF OBJECTS TO BE NOTED, AND SKETCHES.

**236.** The field notes and plat of a survey are designed to furnish not only a technical record of the procedure, but also of equal importance a report upon the character of the land, soil and timber traversed by the survey, and a detailed schedule of the topographical features along every line, with accurate connections showing the relation of the rectangular surveys to other surveys, to natural objects and to improvements. A triple purpose is thus served: (*a*) the technical procedure is made a matter of official record; (*b*) general information relating to a region is gathered; and, (*c*) the "calls" of the field notes and the representations of the plat in respect to objects along the surveyed lines furnish important evidence by which the locus of the survey becomes practically unchangeable as contemplated by law.

The specimen field notes and plats are intended to standardize the form of record, and many special matters relating to these subjects are brought together in Chapters VIII and IX, but before concluding the special questions concerning rectangular surveys it is deemed expedient to outline the technical and topographical features which are to be carefully observed and recorded in the field during the progress of the public-land surveys:—

1. The precise course and length of every line run, noting all necessary offsets therefrom, with the reason for making them, and method employed.

2. The kind and diameter of all bearing trees, with the course and distance of the same from their respective corners, and the markings; all bearing objects and marks thereon, if any; and the precise relative position of witness corners to the true corners.

3. The kind of material of which corners are constructed, their dimensions and markings, depth set in the ground, and their accessories.

4. Trees on line. The name, diameter and distance on line to all trees which it intersects, and their markings.

5. Intersections by line of land objects. The distance at which the line intersects the boundary lines of every reservation, townsite, or private claim, noting the exact bearing of such boundary lines, and the precise distance to the nearest boundary corner; the center line of every railroad, canal, ditch, electric transmission line, or other right-of-way across public lands, noting the width of the right-of-way and the precise bearing of the center line; the change from one character of land to another, with the approximate bearing of the line of demarcation, and the estimated height in feet of the ascents and descents over the principal slopes typifying the topography of the country traversed, with the direction of said slopes; the distance to and the direction of the principal ridges, spurs, divides, rim rock, precipitous cliffs, etc.; the distance to where the line enters or leaves heavy or scattering timber, with the approximate bearing of the margin of all heavy timber, and the distance to where the line enters or leaves dense undergrowth.

6. Intersections by line of water objects. All unmeandered rivers, creeks and smaller water-courses which the line crosses; the distance measured on the true line to the center of the same in the case of the smaller streams, and to both banks in the case of the larger streams, the course downstream at points of intersection, and their widths on line, if only the center is noted. All intermittent water-courses, such as ravines, gulches, arroyos, draws, dry-drains, etc.

7. The land's surface; whether level, rolling, broken, hilly or mountainous.

8. The soil; whether rocky, stony, gravelly, sandy, loam, clay, etc., and also whether first, second, third or fourth rate.

9. Timber; the several kinds of timber and undergrowth, in the order in which they predominate.

10. Bottom lands to be described as upland or swamp and overflowed, as contradistinguished under the law, noting the extent and approximate position of the latter, and depth of overflow at seasonal periods. The segregation of lands fit for cultivation without artificial drainage, from the swamp and overflowed lands, where the latter are subject to selection by the States, is always accomplished by legal subdivision, and any smallest legal subdivision is classified as all upland or all swamp and overflowed land accordingly as more than half of the same may be of the character of the one or of the other class of lands; bottom lands will be classified with special consideration to these matters.

11. Springs of water, whether fresh, saline, or mineral, with the course of the stream flowing therefrom. The location of all streams, springs, or water-holes, which because of their environment may be deemed to be of value in connection with the utilization of public grazing lands, and which may be designated as public watering places, will be specially noted.

12. Lakes and ponds, describing their banks, tributaries and outlet, and whether the water is pure or stagnant, deep or shallow.

13. Improvements; towns and villages; post offices; Indian occupancy; houses or cabins, fields, or other improvements, with owner's name; mineral claims; mill-sites; United States mineral monuments, and all other official monuments not belonging to the system of rectangular surveys; will be located by bearing and distance or by intersecting bearings from given points.

14. Coal banks or beds, all ore bodies, with particular description of the same as to quality and extent; all mining surface improve-

ments and underground workings; and salt licks. All reliable information that can be obtained respecting these objects, whether they be on the line or not, will appear in the general description.

15. Roads and trails, with their directions, whence and whither.

16. Rapids, cataracts, cascades, or falls of water, in their approximate position and estimated height of their fall in feet.

17. Stone quarries and ledges of rocks, with the kind of stone they afford.

18. Natural curiosities, petrifactions, fossils, organic remains, etc.; also all archaeological remains, such as cliff dwellings, mounds, fortifications, or objects of like nature.

19. The general average of the magnetic declination in the township, with maximum known range of local attraction and other variations, will be stated in the general description, and the general average for the township, subject to local attraction, will be shown upon the plat.

20. *General description.*—The above information will be summarized by townships in a general description which will be made the concluding part of the field notes of every survey. The general description will be made to embrace many more comprehensive details in regard to the characteristics of the region than is feasible to cover as an intimate part of the technical record of the survey, as follows:—

*Land.*—A general outline of the drainage and topographical features of the township and approximate range of elevation above sea level.

*Soil.*—The prevailing and characteristic soil types. (See special reference to soil classification, Chap. VII.)

*Timber.*—The predominant forest species, age, size, condition, etc.

*Evidence of mineral.*—All known bodies of mineral, and lands whose formation suggests mineral-bearing characteristics, especially with reference to lands of volcanic or igneous origin, will be listed by appropriate legal subdivision, with brief description of the mineral indications. On the other hand, if the surveyor finds no apparent indication of mineral deposits, a report to that effect will be embodied in the general description.

*Watering places.*—The areas embracing all streams, springs, or water holes as may be of special value as public watering places, in connection with the utilization of public grazing lands, will be listed by appropriate legal subdivision, with brief description of the nature of such water supply.

*Settlement.*—The extent of the settlement at the time of the survey.

*Industry.*—The industrial possibilities of the township, especially as to the adaptability of the region to agricultural pursuits, stock raising, lumbering, mining, or other profitable enterprise.

*Special.*—All exceptional steps in the technical process of the survey, and other special matters required in paragraphs Nos. 1 to 19, inclusive, of the above summary, not otherwise suitably recorded will be reported in the general description.

In addition to the field notes the surveyors are required to prepare, as the work progresses, an outline diagram showing the course and length of all established lines with connections, and a topographical sketch embracing all features usually shown upon the completed official township plat. These maps will be made to scale, drawn in pencil only, if desired, and will be kept up with the progress of the field work. The interiors of the sections will be fully completed; the topographical features will be sketched with care while in the view of the surveyor, and the position within the section of the various details which are to be shown on the completed plat will be located with an accuracy commensurate with their relative importance. The design of the specimen township plat will be followed closely in the preparation of the outline diagram and topographical sketch plat, except that it will generally be desirable to employ a separate sheet for each of the two purposes. These maps will then form the basis of the official plat, the ultimate purpose of which is a true and complete graphic representation of the public lands surveyed.

# CHAPTER IV.

## CORNER MONUMENTS.

### THE LEGAL SIGNIFICANCE OF A CORNER MONUMENT.

**237.** It is one of the fundamental principles of the surveying laws that absolute permanency be attached to the public-land surveys when the lines have been officially established. The "survey" embraces certain definite technical procedure, heretofore described, also the marking of certain fixed points, as will be described in this chapter, though the establishment of a survey may not be termed "completed" until the field notes and plat and every detail of the technical operation constituting the survey have been finally accepted by the Commissioner of the General Land Office, all as contemplated by law. The law provides that the original corners established during the process of the survey shall forever remain fixed in position, even to disregarding technical errors in the execution of the survey—where discrepancies may have passed undetected prior to the acceptance of the survey and the opening of the lands to entry—and, as an aid to the matter of permanency, the Congress provides for the purchase of durable material for the corner monuments, also a penalty for the defacing of any marks relating to the locus of the survey. If it were possible to carry out the full intent of the surveying laws in regard to the aforementioned particulars, the most intricate of all technical and legal problems relating to surveys—the questions pertaining to the reestablishment of lost corners—would be avoided.

The courts attach major importance to authentic evidence relating to the original position of an official corner monument, such evidence being given far greater weight than the technical record relating to bearings and lengths of lines, and it is assumed in the first instance that the original corners shall serve every necessary purpose for the identification of the survey delineated upon the official approved plat, and of the lands which have passed into private ownership. The legal significance of the original monuments, as thus briefly outlined, makes it mandatory upon the surveyor to exercise constant diligence in the workmanlike construction of lasting corners, and alertness in skillfully connecting the same with natural objects or improvements, to the end that the greatest possible permanency may be secured for the public-land surveys.

**238.** Accordingly, if a surveyor is called upon to alter the condition of a previously established point, the utmost regard should be shown for the evidence of the original location of the monument, and the corner will be carefully reconstructed by such additional means as may be appropriate, without destroying the evidence which served to identify its legal position. A complete record will be kept of the description of the old monument as identified, and all alterations and additions thereto.

**239.** Regulation monuments are employed to mark permanently the position of the quarter-section, section, township and meander corners, appropriate to the subdivision of the public lands, as described in Chapter III; also at such sixteenth-section corners as the requirements of the written special instructions or the exigencies of the survey of fractional sections may demand; also at all angle points along an irregular boundary line, and at intermediate intervals of 40 and 80 chains along such limiting boundary. A more extended discussion of the subject of "angle points" and other monuments to be established upon irregular boundaries will be found in Chapter VII.

**240.** The position of every corner monument will be "evidenced" by the best of such accessories as may be available, and where the corner point itself can not be marked in the usual manner an appropriate "witness corner" will be established. A "witness meander corner" will be established upon secure ground wherever the intersection of a surveyed line with the mean high-water elevation of a meanderable body of water falls at a point where the monument would be liable to destruction.

**241.** The field notes relating to the establishment of a corner monument will be introduced into the technical record of the survey at the logical place in the record where the true position for the corner is indicated as having been attained. The record of the monument itself will embrace a description of:

(*a*) The corner material, including its dimensions, in the order of length and diameter of an iron post; or length, width and breadth of a stone; or the breast height diameter of a tree; (*b*) the depth set in the ground, with mention of additional support if any; (*c*) the significance of its position; (*d*) the markings upon the monument; and (*e*) the nature of the accessories, including character, size, position and markings.

### CORNER MATERIAL.

**242.** The General Land Office has adopted a model iron post for monumenting the public-land surveys, which will be generally used unless exceptional circumstances warrant a departure from this rule. This practice is deemed so important that the surveyor is not authorized to exercise an option in the matter, but he may refer the question to the proper supervising officer, who may grant authority for the use of other suitable material, provided the reasons for departing from the general rule are sufficient, in which case a brief statement of the facts will be given in the field notes, in the form of an explanation as to why the model iron posts were not employed.

The model iron post is made from commercial iron pipe, from 1 to 3 inches in diameter, which is cut into lengths of about 36 inches; one end of the pipe is split for a distance of about 4 or 5 inches, and the two halves are spread (when heated) to form flanges or foot plates, at right angles to the axis of the pipe; a brass cap is securely riveted to the opposite end of the pipe; and finally the pipe is filled with concrete. Unless otherwise provided in the written special instructions, the iron posts will be employed as follows: 3-inch, for standard and closing township corners, corners of one, two or four townships, and as required for mile corners and angle points of special boundary surveys; 2-inch, for standard and closing section corners, and corners of one, two or four sections; and, 1-inch, for quarter-section and meander corners, and as required for miscellaneous angle points, sixteenth-section corners and corners of special tract surveys. All witness corners are to be of the same size as would be used for the true corner.

**243.** The caps of the iron posts are to be suitably and plainly marked with steel dies at the time when used; the posts will be set in the ground about three-fourths of their length; and earth and stone, if the latter is at hand, will be tamped into the excavation to give the post a solid anchorage.

**244.** Durable native stone may be substituted for the model iron post, if the procedure has been duly authorized, but no stone will be used which measures less than 20 inches in length, or less than 6 inches in either of its minor dimensions, or less than 1,000 cubic inches in volume. A stone should always be selected with regard to its durability when exposed to the usual weathering influences. Stone will not be used as a corner monument where its position falls among large quantities of loose surface stone or slide rock.

**245.** A stone will be suitably and legibly marked with a steel chisel or punch with such letters, figures, grooves or notches, as may be required, and will be set firmly in the ground about three-fourths of its length.

**246.** Both iron post and stone monuments will always be set the usual depth in the ground unless it is impossible to complete the excavation, in which case the monument will be planted as deep as conditions will permit, and the necessary support will be secured by a stone mound.

**247.** Where the corner point falls upon solid surface rock, preventing excavation, a cross (X) will be cut at the exact corner point, and, if feasible, the monument will be erected in the same position, supported by a large stone mound of broad base, so well constructed that it will possess thorough stability.

**248.** Where the corner point falls exactly at the position occupied by a sound living tree, which is too large to be removed, the tree will be appropriately marked for the corner.

### WITNESS CORNERS.

**249.** Where the true point for a corner falls within a roadway in such a place as to interfere with travel, a marked (X) stone will be deposited in the ground at the true corner point and a witness corner will be established at some suitable point, preferably on a surveyed line, outside of the roadway.

**250.** Where the true point for a corner falls upon insecure ground, or in an inaccessible place, such as within an unmeandered stream, lake or pond, or in a marsh, or upon a precipitous slope or cliff, a witness corner will be established at some suitable point, preferably on a surveyed line, where the monument may be permanently constructed.

**251.** The surveyor will be expected to exercise his best judgment in selecting the position for a witness corner, with a view to affording a definite and convenient connection from the witness corner to the true point for the monument, for use in subsequent surveys to recover the legal position of the true corner. Extra effort will be exerted to accomplish the permanent establishment of a monument at its true corner point, wherever this is feasible, in order to avoid as much as possible the confusion to settlers and others caused by witness corners.

**252.** Only one witness corner will be established in each instance, and the same will be placed upon any one of the surveyed lines leading to a corner, if a suitable place, within a distance of 10 chains, is available, but if there is no secure place to be found on a surveyed line within the stated limiting distance, the witness corner may be located in any direction within a distance of 5 chains. On the other hand, if there is no suitable place within the latter radius, one or more legal subdivisions will be eliminated from the survey as provided in Chapter VII.

**253.** All of the lines of a survey will be completed in the regular manner, if the true point for a corner is accessible, but where the true point can not be attained, a line connecting therewith may be returned as surveyed if the same has been completed by the projection and measurement of a suitable offset or traverse, resulting in a closed figure which approaches the true point for a monument within the limit prescribed for the establishment of witness corners.

**254.** The field notes will show every detail of the relation of a witness corner to the true point for a monument, and the direct connecting course and distance will be shown upon the plat of the survey.

### MARKING CORNERS.

**255.** All classes of corner monuments are to be marked in accordance with a system hereinafter described which has been devised to furnish a ready identification of the character and position of the monument which bears the marks. Capital letters and Arabic figures are employed to mark iron post and tree corners, while upon stone corners certain additional marks termed "notches" and "grooves" are employed to convey the same information, but to lessen the labor incident to the marking process. The letters and figures upon a monument are designed to relate to the township, range and section to which the corner belongs; the notches and grooves upon a stone monument relate—in the case of an exterior corner—to the normal number of miles from the monument to the adjoining township corners, and—in the case of a subdivisional corner—to the normal number of miles from the monument to the township boundary lines, as hereinafter described, thus furnishing the means of ascertaining the appropriate section numbers.

**256.** All markings should be accomplished neatly, distinctly and durably; and the marks are to be carefully arranged. An assortment of steel dies, chisels, punches and timber scribes, in perfect condition for use, should always be at hand.

**257.** A witness corner and its accessories will be constructed and marked similarly to a regular corner for which it stands, with the additional letters "W C" to signify "witness corner."

**258.** The following schedule is an index of the ordinary markings common to all classes of corners and accessories:—

| Marks. | To indicate. | Marks. | To indicate. |
|---|---|---|---|
| A M C | Auxiliary meander corner. | R | Range. |
| | | S | Section. |
| A P | Angle point. | S | South. |
| B O | Bearing object. | S C | Standard corner. |
| B T | Bearing tree. | SE | Southeast. |
| C | Center. | S M C | Special meander corner. |
| C C | Closing corner. | SW | Southwest. |
| E | East. | T | Township. |
| M | Mile. | TR | Tract. |
| M C | Meander corner. | W | West. |
| N | North. | W C | Witness corner. |
| NE | Northeast. | W P | Witness point. |
| NW | Northwest. | $\frac{1}{4}$ | Quarter section. |
| P L | Public land (unsurveyed). | $\frac{1}{16}$ | Sixteenth section. |

MARKS ON IRON POST MONUMENTS.

**259.** The markings upon the brass cap of an iron post should always be made to read from the south side of the monument, and all iron posts will be marked with the year number at the date when established.

**260.** *Standard township corners* are to be marked "S C" and the township on the north half, and the ranges and sections in the proper quadrants; as for example:

SC
T 25 N
R 17 E | R 18 E
S 36 | S 31
1916

**261.** *Closing township corners* are to be marked "C C" on the half from which the closing line approaches the monument, with the township (or range) on the same half, and the ranges (or townships) and sections in the proper quadrants; also (as far as known at the time) the township, range and section, or the initials or abbreviation of the State, reservation, grant or private claim, upon which the township exterior closes; as for example:

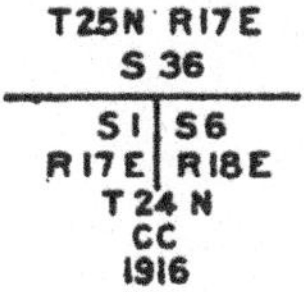


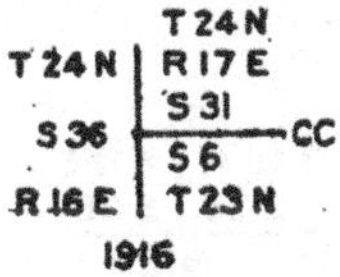


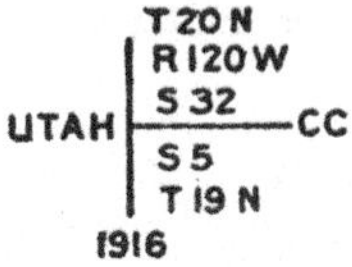


**262.** *Corners common to four townships* are to be marked with the townships on the north and south halves, the ranges on the east and west halves, and the sections in the four quadrants; as for example:

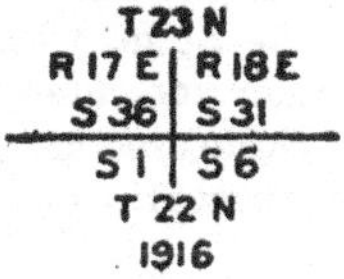


**263.** *Corners common to two townships only* are to be marked with the township (or range) common to both on the proper half, and the ranges (or townships) and sections in the proper quadrants; also (as far as known at the time) the township, range and section upon the opposite half; as for example:

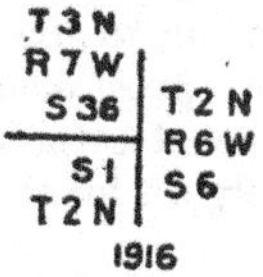


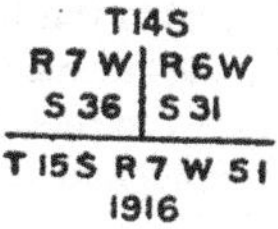


**264.** *Corners referring to one township only* are to be marked with the township, range and section in the particular quadrant which is concerned; also (as far as known at the time) the township, range and section upon the opposite part; as for example:

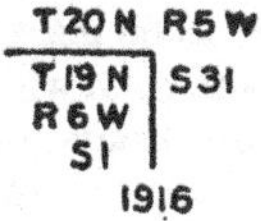

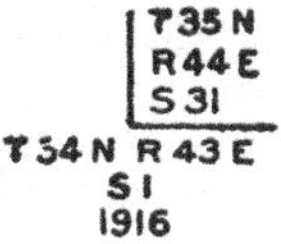


55465°—19——16

**265.** *Standard section corners* are to be marked "S C" and the township and range on the north half, and the sections in the proper quadrants; as for example:

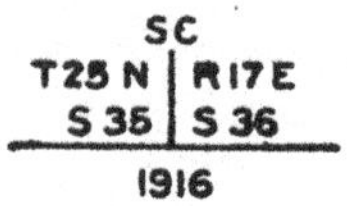


**266.** *Closing section corners* are to be marked "C C" and the township and range on the half from which the closing line approaches the monument, and the sections in the proper quadrants; also (as far as known at the time) the township, range and section, or the initials or abbreviation of the State, reservation, grant or private claim, upon which the section line closes, with the exception that in the case of an interior closing section corner, the township and range numbers will not be repeated; as for example:

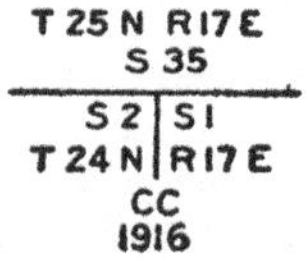


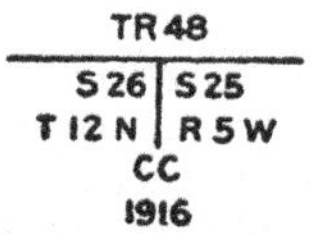


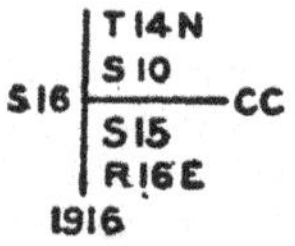


**267.** *Corners common to four sections* are to be marked: (*a*) On an exterior, with the township (or range) common to the adjoining townships, the ranges (or townships) upon the opposite sides of the exterior, and the sections; and (*b*) a subdivisional corner, with the township, range and sections; all appropriately set forth as follows:

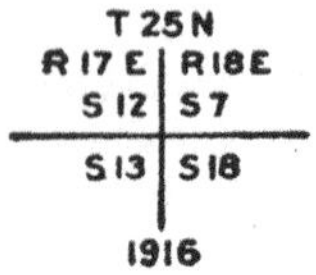


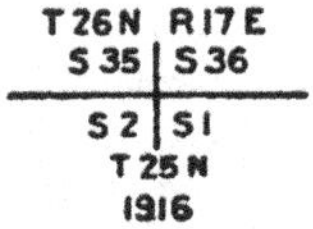


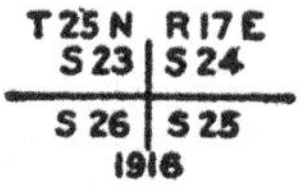


**268.** *Section corners common to two sections only* are to be marked with the township and range on the half facing the sections to which the corner belongs, and the sections in the proper quadrants; also (as far as known at the time) the township, range and section upon the opposite half, except that in the case of an interior corner, the township and range numbers will not be repeated; as for example:

```
T14S | T14S
 S12 |
-----| R18E
 S13 |
R17E | S7
   1916
```

```
T27N | R17W
 S31 | S32
------------
T26N R17W
   S 6
   1916
```

```
T14S | R20W
 S10 | S11
------------
   S14
   1916
```

**269.** *Section corners referring to one section only* are to be marked with the township, range and section in the particular quadrant which is concerned; also (if known at the time) the section upon the opposite part; as for example:

```
     S10
 ------
 T84N |
 R73W |
  S16 |
     1916
```

```
T27N |
R16W |
 S17 |
------
     S20
     1916
```

```
 S28
  ------
  | T57N
  | R63W
  | S34
 1916
```

**270.** *Standard quarter-section corners* are to be marked "S C ¼" and the section, all on the north half; as for example:

```
   SC
 ¼S 36
--------
  1916
```

**271.** *Quarter-section corners of maximum control* are to be marked (*a*) on a meridional line, "¼" on the north, and the sections on the east and west halves; and, (*b*) on a latitudinal line, "¼" on the west, and the sections on the north and south halves; as for example:

```
     ¼
S 13 | S 18
     |
   1916
```

```
    S 21
¼  ------
    S 28
    1916
```

**272.** *Quarter-section corners of minimum control* are to be marked "¼" and the section, all on the half toward the particular section which is concerned; as for example:

```
-------
¼ S 4
 1916
```

```
¼ S 16
-------
 1916
```

```
| ¼S7
|
 1916
```

**273.** *Meander corners* are to be marked "M C" on the half toward the meanderable body of water, and the additional marks (*a*) on a standard parallel or other line controlling surveys to one side only, with the township, range and section toward the surveyed land; (*b*) on an exterior, with the township (or range) common to the adjoining townships, the ranges (or townships) upon the opposite sides of the exterior, and the sections; and, (*c*) on a subdivisional line, with the township, range and sections; all appropriately set forth as follows:

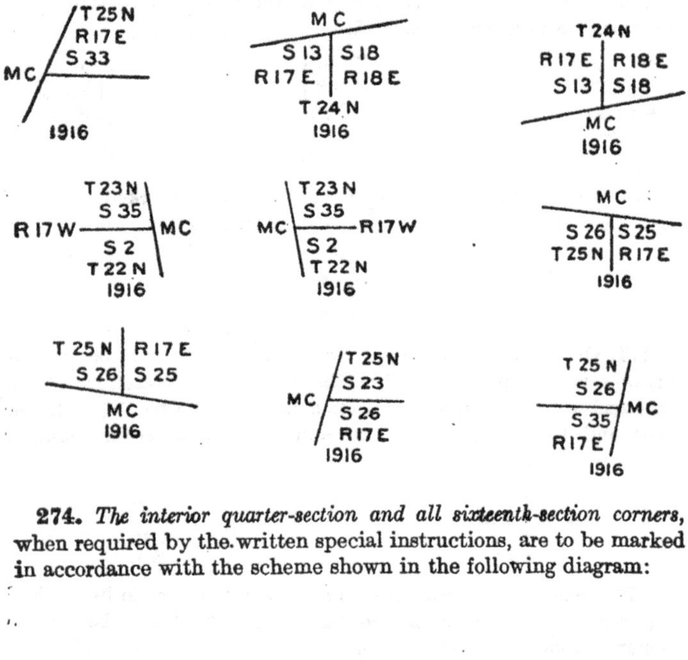


**274.** *The interior quarter-section and all sixteenth-section corners,* when required by the written special instructions, are to be marked in accordance with the scheme shown in the following diagram:

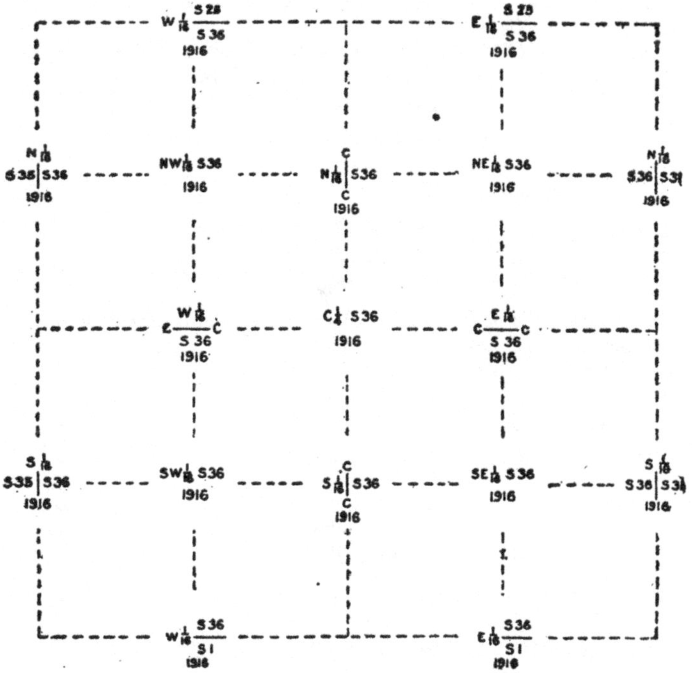


**275.** *Sixteenth-section corners of minimum control* are to be marked with a key letter (N, E, S or W), to indicate the position of the monument, and "$\frac{1}{16}$" and the section, all on the half toward the particular section which is concerned; as for example:

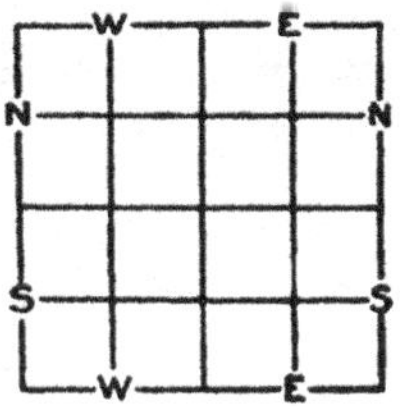


Key

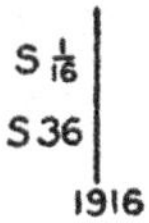

**276.** *Special meander corners* are to be marked in accordance with the following scheme:

Key letters (N, E, S, W or C) will be used in pairs to indicate the position of the subdivision - of - section line.

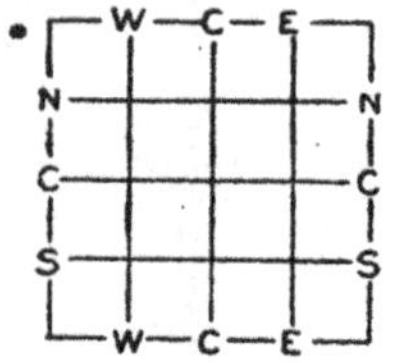


Key

The marks "S M C" will be placed on the half toward the meanderable body of water, and the section on the opposite half, as for example:

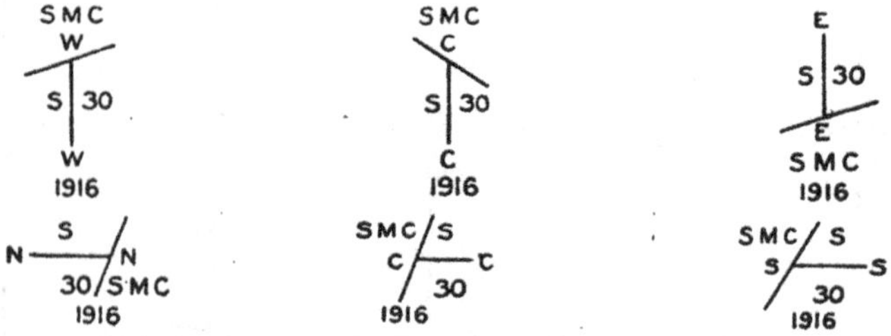


**277.** *Auxiliary meander corners* will be marked "A M C" and the township, range and section; as for example:

A M C
T 64 N R 37 W
S 29
1916

**278.** *Closing subdivision - of - section corners* are to be marked in accordance with the following scheme:

Key letters (N, E, S, W or C) will be used in pairs to indicate the position of the subdivision - of - section line.

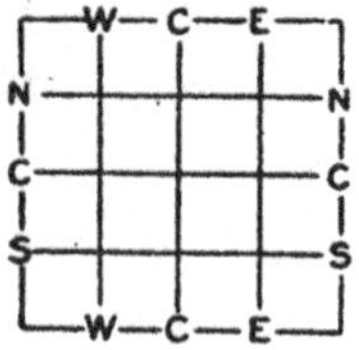


Key

The marks "C C" and the section will be placed on the half from which the closing line approaches the monument.

(The marks "B I R" indicate "Blackfeet Indian Reservation.")

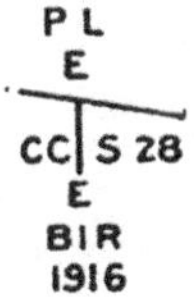


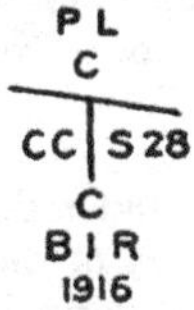

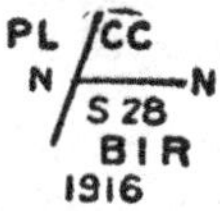

**279.** Markings for miscellaneous angle points along irregular boundaries:

For "angle point No. 4" on the boundary of the "Blackfeet Indian Reservation," falling on surveyed land.

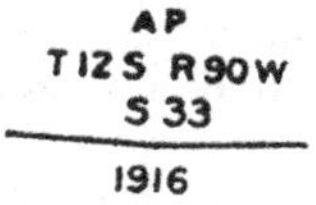


For "angle point" on the south boundary of section 33, superseding an old standard corner on a defective line, not subject to rectification.

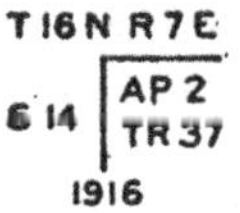


For "angle point No. 2" on the boundary of a private claim ("Tract No. 37") falling on surveyed land.

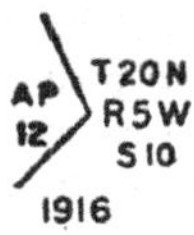


For "angle point No. 12" on a reestablished meander line; the marks "A P" and the serial number will be placed on the half toward the land omitted from the original survey.

**280.** Markings for intermediate corners along irregular boundaries:

For "139th mile corner" on the boundary line between the States of "New Mexico and Texas."

For "3d mile corner" on the boundary of the "Blackfeet Indian Reservation," falling on unsurveyed land.

For "13th mile corner" on the boundary of the "Blackfeet Indian Reservation," falling on surveyed land.

### MARKS ON STONE MONUMENTS.

**281.** Where a stone monument is established the letters, figures, and grooves will be cut on the exposed faces or sides of the stone, but not on its top or end; the notches will be cut upon the exposed vertical edges. Grooves are employed where the faces of a stone are oriented to the cardinal directions, and notches where the vertical edges are turned to the cardinal points. All marks will be made from 1 to 1½ inches in size, and will be plainly and permanently chiseled into the stone.

**282.** *Standard township corners* (oriented with the faces to the cardinal directions) are to be marked "S C" on the north face, with the township on the same face, and the ranges on the adjoining faces; as for example:

S C 25 N on N.,
18 E " E., and
17 E " W. face.

**283.** *Closing township corners* (oriented with the faces to the cardinal directions) are to be marked "C C" and with six (or fewer) grooves on the face from which the closing line approaches the monument—the grooves to indicate the normal number of miles (or fractional parts) from the monument to the adjoining township corner—with the township (or range) on the same face, and the ranges (or townships) on the adjoining faces; also the initials or abbreviation of the State, reservation, grant or private claim, on the face toward such irregular tract as may be closed upon; as for example:

20 N on N.,
C C 120 W and 5 grooves (on line between sections 5 and 32) on E.,
19 N on S., and
UTAH " W. face.

**284.** *Corners common to four townships* (oriented with the edges to the cardinal points) are to be marked with the townships on the northeast and southwest faces, and the ranges on the southeast and northwest faces; as for example:

23 N on NE.,
18 E " SE.,
22 N " SW., and
17 E " NW. face.

**285.** *Corners common to two townships only* (oriented with the faces to the cardinal directions) are to be marked with the township (or range) common to both on the face toward the townships, and the ranges (or townships) on the adjoining faces; as for example:

3 N on N.,
2 N " S., and
7 W " W. face.

**286.** *Corners referring to one township only* (oriented with the edges to the cardinal points) are to be marked with the township and range on the face toward the particular township; as for example:

23 N 7 W on NW. face.

**287.** *Standard section corners* (oriented with the faces to the cardinal directions) are to be marked "S C" on the north face, and with from one to five grooves on the east and west faces, the grooves to indicate, respectively, the number of miles from the monument to the adjoining (regular) township corner; as for example:

S C on N.,
1 groove on E., and
5 grooves on W. face (standard corner of sections 35 and 36).

**288.** *Closing section corners* (oriented with the faces to the cardinal directions) are to be marked "C C" and with from one to six grooves on the face from which the closing line approaches the monument, and from one to five grooves on each of the adjoining faces—the grooves to indicate the number of miles (or fractional parts) from the monument to each of the three (regular) township boundary lines in the same directions, respectively—also the initials or abbreviation of the State, reservation, grant or private claim, on the face toward such irregular tract as may be closed upon; as for example:

2 grooves on E.,
C C and 6 " " S., and
4 " " W. face (on line between sections 2 and 3 closing on a standard parallel).

**289.** *Corners common to four sections* (oriented with the edges to the cardinal points) are to be marked (*a*) on an exterior, with from one to five notches each on two opposite edges, north and south on a meridional line, and east and west on a latitudinal line, each to indicate, respectively, the number of miles from the monument to the adjoining (regular) township corner; and (*b*) a subdivisional corner, with from one to five notches on the east and south edges, each to indicate, respectively, the number of miles from the monument to the (regular) east and south township boundary lines; the subdivisional section corners of a fractional township will be marked with reference to the theoretical position of normal east and south boundaries, whether surveyed or not; as for example:

2 notches on N. and 4 notches on S. edge (for corner of sections 7, 12, 13 and 18 on a range line).

2 notches on E. and 4 notches on W. edge (for corner of sections 2, 3, 34 and 35 on a township line).

2 notches on E. and 4 notches on S. edge (for corner of sections 10, 11, 14 and 15 of a subdivisional survey).

**290.** *Section corners common to two sections only* (oriented with the edges to the cardinal points) are to be marked with the sections on the faces toward the particular sections to which the corner belongs; as for example:

S 13 on SW., and
S 12 " NW. face (for corner of sections 12 and 13 on the east boundary of a township).

S 11 on NE., and
S 10 " NW. face (for corner of sections 10 and 11 of a subdivisional survey running north from the monument).

**291.** *Section corners referring to one section only* (oriented with the edges to the cardinal points) are to be marked with the section on the face toward the particular section which is concerned; as for example:

S 17 on NW. face (for southeast corner of section 17).

**292.** *Standard quarter-section corners* (oriented with the faces to the cardinal directions) are to be marked "S C ¼" on the north face.

**293.** *Quarter-section corners of maximum control* (oriented with the faces to the cardinal directions) are to be marked (*a*) on a meridional line, "¼" on the west face; and (*b*) on a latitudinal line, "¼" on the north face.

**294.** *Quarter-section corners of minimum control* (oriented with the faces to the cardinal directions) are to be marked "¼" and the section, all on the face toward the particular section which is concerned; as for example:

¼ S 4 on S. face (for quarter-section corner on the north boundary of section 4).

**295.** *Meander corners* (oriented with the faces to the cardinal directions) are to be marked "M C" on the face toward the meanderable body of water, and with from one to six grooves on each of the other faces, each to indicate the number of miles (or fractional parts) from the monument to the (regular) township boundary line in the same direction, respectively; as for example:

M C on N.,
6 grooves " E.,
4 " " S., and
6 " " W. face (for meander corner of fractional sections 13 and 18, on the south side of a meanderable body of water).

**296.** *Special and auxiliary meander corners* (oriented with the faces to the cardinal directions) are to be marked "S M C" or "A M C," as the case may be, on the face toward the meanderable body of water, and the section on the opposite face; as for example:

S M C on N., and
S 19 " S. face (for special meander corner on a meridional subdivision-of-section line in section 19, on the south side of a meanderable body of water).

S 20 on E., and
A M C " W. face (for auxiliary meander corner in section 20, on the east side of a meanderable body of water).

MARKS ON TREE MONUMENTS.

**297.** Where the true point for a corner is found to fall in the position occupied by a sound living tree, which is too large to be removed, the tree will be made the monument. A tree will be removed if it is too small to be marked, and a witness corner will be established in preference to marking an unsound tree, if the latter can not be removed.

**298.** The species of the tree and its diameter, breast height, will be noted, where a tree is to be made a monument, and the appropriate marks will be made upon the trunk of the tree immediately above the root crown. A series of marks to be made upon a particular side of a tree will be scribed in a vertical line reading downward.

**299.** In the case of certain trees, including the aspen, beech and locust (smooth, thin and permanently barked from sapling to maturity), the marks may be made preferably by scribing well into the bark and cambium (or live wood tissue) without blazing; the marks thus made will remain and be visible as long as the tree is sound; on the other hand, in the case of practically all rough barked trees, the marks should be scribed into a smooth, narrow, vertical blaze, specially prepared by removing just enough of the outer growth to expose a flat surface of the live wood tissue immediately underneath the bark; the marks thus made will remain as long as the tree is sound, but the blaze and marks will be covered by a gradual overgrowth, showing an outward scar for many years. In regions subject to heavy snowfall it is desirable to make a small additional blaze at a height of 6 or 8 feet above the ground, which will serve to attract attention to the tree during the winter season. The ends of all blazes should be smoothed off gradually without making a sharp cut into the cambium. The lower end of the blaze upon which the marks are placed should be about 6 inches above the root crown, and its length should be just sufficient to take the marks.

The practice relating to the manner of marking trees, as above outlined, is designed to cause the least possible injury to the tree, by enabling a rapid overgrowth; also, to place the marks in a position where they will remain on the stump if the trunk should be removed. Various practices have obtained in the past in different localities, some of which are objectionable by causing unnecessary injury to a tree, or on account of the marks being placed in a position where there is danger of their removal with the trunk in case the tree is cut down.

**300.** The above theory applies equally to the marking of bearing trees, and the surveyor is advised, when making retracements, resurveys, etc., not to remove the overgrowth on a tree monument or bearing tree unless it is absolutely necessary to do so in order to identify positively the particular tree. In the case of trees which have been blazed before marking, the number of rings contained in the overgrowth (or its equivalent on the adjoining section of the tree) will furnish an exact count of the number of years (one annual ring for each growing season) from the date of original marking to the date when uncovered. After an old blaze has been uncovered, conditions are favorable for the decaying process to set in, and the surveyor should adopt additional means to evidence the position of the corner.

**301.** *Standard township corners* are to be marked "S C" and the township on the north side, and the ranges and sections on the east and west sides; as for example:

S C T 25 N on N.,
R 18 E S 31 " E., and
R 17 E S 30 " W. side.

**302.** *Closing township corners* are to be marked "C C" and the township (or range) on the side from which the closing line approaches the monument, and the ranges (or townships) and sections on the adjoining sides; also the initials or abbreviation of the State, reservation, grant or private claim, on the side toward any irregular tract which may be closed upon; as for example:

R 18 E S 6 on E.,
C C T 24 N " S., and
R 17 E S 1 " W. side.

**803.** *Corners common to four townships* are to be marked with the township and section on the northeast and southwest sides, and the range and section on the southeast and northwest sides; as for example:

T 23 N S 31 on NE.,
R 18 E S 6 " SE.,
T 22 N S 1 " SW., and
R 17 E S 36 " NW. side.

**804.** *Corners common to two townships only* are to be marked with the township, range and section on the sides toward the particular townships; as for example:

T 2 N R 7 W S 1 on SW., and
T 3 N R 7 W S 36 " NW. side.

**805.** *Corners referring to one township only* are to be marked with the township, range, and section on the side toward the particular township which is concerned; as for example:

T 23 N R 7 W S 36 on NW. side.

**806.** *Standard section corners* are to be marked "S C" and the township and range on the north side, and the sections on the east and west sides; as for example:

S C T 25 N R 17 E on N.,
S 36 " E., and
S 35 " W. side.

**807.** *Closing section corners* are to be marked "C C" and the township and range on the side from which the closing line approaches the monument, and the sections on the adjoining sides; also the initials or abbreviation of the State, reservation, grant or private claim on the side toward any irregular tract which may be closed upon; as for example:

S 1 on E.,
C C T 24 N R 17 E " S., and
S 2 " W. side.

**808.** *Corners common to four sections* are to be marked (*a*) on an exterior, with the township (or townships), ranges (or range) and sections; and (*b*) a subdivisional corner, with the township, range and section; all appropriately set forth as follows:

T 25 N S 7 on NE.,
R 18 E S 18 " SE.,
R 17 E S 13 " SW., and
S 12 " NW. side.

T 26 N S 36 on NE.,
R 17 E S 1 " SE.,
T 25 N S 2 " SW., and
S 35 " NW. side.

T 25 N S 24 on NE.,
R 17 E S 25 " SE.,
S 26 " SW., and
S 23 " NW. side.

**809.** *Section corners common to two sections only* are to be marked with the township and section and the range and section on the sides toward the particular sections to which the corner belongs; as for example:

T 14 S S 11 on NE., and
R 20 W S 10 " NW. side.

**810.** *Section corners referring to one section only* are to be marked with the township, range and section on the side toward the particular section which is concerned; as for example:

T 27 N R 16 W S 17 on NW. side.

**811.** *Standard quarter-section corners* are to be marked "S C ¼" and the section, all on the north side; as for example:

S C ¼ S 36 on N. side.

**812.** *Quarter-section corners of maximum control* are to be marked (*a*) on a meridional line, "¼" and the section on the west side, and the section on the east side; and (*b*) on a latitudinal line, "¼" and the section on the north side, and the section on the south side; as for example:

S 18 on E., and
¼ S 13 " W. side.

¼ S 21 on N., and
S 28 " S. side.

**813.** *Quarter-section corners of minimum control* are to be marked "¼" and the section, all on the side toward the particular section which is concerned; as for example:

¼ S 7 on E. side (for quarter-section corner on the west boundary of section 7).

**814.** *Meander corners* are to be marked "M C" on the side toward the meanderable body of water, and the additional marks (*a*) on a standard parallel or other line controlling surveys to one side only, with the township, range and section on the side toward the surveyed land; (*b*) on an exterior, with the township (or range) common to the adjoining townships on the side opposite the meanderable body of water, and the ranges (or townships) and the sections on the adjoining sides; and, (*c*) on a subdivisional line, with the township and range on the side opposite the meanderable body of water, and the sections on the adjoining sides; as for example:

M C on E., and<br>
T 25 N R 17 E S 33 " NW. side (for meander corner on a standard parallel, on the west side of a meanderable body of water).

T 24 N on N.,<br>
R 18 E S 18 " E.,<br>
M C " S., and<br>
R 17 E S 13 " W. side (for meander corner on a range line, on the north side of a meanderable body of water).

T 23 N S 35 on N.,<br>
M C " E.,<br>
T 22 N S 2 " S., and<br>
R 17 W " W. side (for meander corner on a township line, on the west side of a meanderable body of water).

S 23 on N.,<br>
T 25 N R 17 E " E.,<br>
S 26 " S., and<br>
M C " W. side (for meander corner on a latitudinal section line, on the east side of a meanderable body of water).

M C on N.,<br>
S 9 " E.,<br>
T 4 N R 7 W " S., and<br>
S 8 " W. side (for meander corner on a meridional section line, on the south side of a meanderable body of water).

**815.** *Special and auxiliary meander corners* are to be marked "S M C" or "A M C", as the case may be, on the side toward the meanderable body of water, and the section on the opposite side; as for example:

S M C on E., and<br>
S 14 " W. side (for special meander corner on a latitudinal subdivision-of-section line in section 14, on the west side of a meanderable body of water).

A M C on N., and<br>
S 9 " S. side (for auxiliary meander corner in section 9, on the south side of a meanderable body of water).

### CORNER ACCESSORIES.

**316.** The purpose of a corner accessory is to evidence the position of the original monument. A connection is made from the monument to fixed natural or artificial objects in its immediate vicinity, whereby the former may be relocated from the latter, thus in the event of the destruction or removal of the corner monument, its original position may be identified as long as any part of the accessories remains in evidence. The accessories consist of three general classes, one or more of which are to be employed at each and every corner established in the public-land surveys, preference being given to the same in the order of their permanency conditional upon the character of the ground in the locality of the monument, as follows:

(*a*) Bearing trees, or other natural objects such as notable cliffs and boulders; permanent improvements; and memorials; (*b*) mound of stone; and (*c*) pits.

**317.** The surveyor can not perform any more important service in connection with his official duties than to employ whatever means may be necessary permanently and accurately to evidence the location of the legal corners established in his survey, and where the usual accessories, or combinations of the same, can not be employed, such other means should be adopted as will best serve the purpose.

**318.** The accessories for witness corners will be the same as though the corner were established at its true point, but the marks upon the bearing trees or other objects will be preceded by the letters "W C", and the section number will be made to agree with the section in which the tree or object actually stands.

### BEARING TREES, BEARING OBJECTS, AND MEMORIALS.

**319.** Bearing trees, or other natural objects, are to be selected for marking when the same are available within a distance of 5 chains of the corner monument, and where the regular quota, hereinafter described, is not available, one tree or object will be marked in each section affording such accessory. A full description of the tree or object will be embodied in the field notes as a part of the record of the corner monument. One tree, or object, will be marked in each section cornering at the monument, when available, and the true course and horizontal distance from the exact corner point to the center vertical axis of the tree at its root crown, or to the cross (X) upon a marked object, will be carefully determined and recorded with the description of the tree, or object, and its marks. The species of a tree and its diameter, at breast height, will be recorded; and, in the case of a cliff or bowlder, the description will embrace such essential details as may be necessary to serve for its ready identification.

**320.** The marks upon a bearing tree will be made upon the side facing the corner and will be scribed in the manner already outlined for marking tree corner monuments. The marks will embrace the information suggested in the schedule hereinafter given, with such letters and figures as may be appropriate for a particular corner, and will include the letters "B T"; a tree will always be marked to agree with the section in which it stands, and will be marked in a vertical line reading downward, ending in the letters "B T" at the lower end of the blaze approximately 6 inches above the root crown.

**321.** There is a great difference in the longevity of trees, and in their rate of decay, etc.; trees should therefore be selected, if possible, with a view to the length of their probable life, their soundness, favorable site conditions and size. Sound trees from 6 to 8 inches in diameter, of the most hardy species, favorably located, are to be preferred for marking. Trees less than 4 inches in diameter will not be selected for marking if larger trees are available, and it is generally better to avoid marking fully matured trees, especially those showing signs of decay. Trees less than 4 inches in diameter, if no better trees are available, will be marked with the letters "B T" only. The species, size and exact position of the bearing trees are of vital importance, as this data will generally serve to identify a bearing tree without uncovering the marks, or even to identify two or more stumps after all evidence of the marks has disappeared.

**322.** Generally only one tree will be marked in each section at a particular corner, but in certain instances, hereinafter described, two trees are required in a section. In such cases it is better to select trees of different species, or of widely different size, direction or distance, if the trees are of the same species, in order that confusion may be avoided in the future identification of a remaining tree where the companion tree has disappeared.

**323.** A cross (×) and the letters "B O" and the section number will be chiseled into a bearing object, if it is of rock formation, and the record should be such as to enable another surveyor to determine where the marks will be found.

**324.** A connection to any permanent artificial object or improvement may be included in this general class of corner accessories. The field notes should be explicit in describing such objects, and should indicate the exact point to which a connection is made, as "southwest corner of foundation of Smith's house," "center of Smith's well," "pipe of Smith's windmill," etc. No marks will be made upon private property.

**325.** In every case where it is impossible to make a single connection to a bearing tree or other bearing object, as above described, and where a mound of stone or pits are impracticable, a suitable memorial will be deposited at the base of the monument. A memorial may consist of any durable article which will serve to identify the location of the corner in case the monument is destroyed. Such articles as glassware, stoneware, a marked (×) stone, a charred stake, a quart of charcoal, or pieces of metal will constitute a suitable memorial. A full description of such articles will be embodied in the field notes wherever they are employed as a corner accessory.

### MOUND OF STONE.

**326.** Where native stone is available and the surface of the ground is favorable, a mound of stone will be employed as an accessory to a corner monument, provided that a full quota of trees or other bearing objects can not be utilized. A mound of stone erected as a corner accessory will be built as stably as possible, will consist of not fewer than five stones, and will be not less than 2 feet base and 1½ feet high. In stony ground the size of the mound will be sufficiently increased to make it conspicuous. The position of the mound will be as shown in the schedule hereinafter stated, and the nearest point on its base will be separated about 6 inches distant from the monument. The field notes will show the size and position of the mound.

**327.** Where it is necessary to support a monument in a stone mound, no additional mound will be employed as an accessory; and, if bearing trees or other objects are not available, a marked (X) stone or other memorial will be deposited at the base of the monument.

PITS.

**328.** Where the full quota of trees or other bearing objects are unavailable for marking, the position of the monument will, under certain favorable conditions, be evidenced by pits. No pits should be dug in a roadway, or where the ground is overflowed for any considerable period, or upon steep slopes, or where the earth will wash, or in a loose or light soil, or where there is no native sod, or where suitable stone for a mound is at hand.

A firm soil covered with a healthy native sod is most favorable for a permanent pit. Under such conditions the pits will gradually fill with a material slightly different from the original soil, and a new species of vegetation will generally take the place of the native grass; these characteristics, under favorable conditions, make it possible to identify the original location of the pits after the lapse of many years.

**329.** All pits will be dug 18 inches square and 12 inches deep, with the nearest side 3 feet distant from the corner monument, oriented with a square side (and not a corner) towards the monument, arranged as shown in the schedule hereinafter given; the earth removed will be scattered in such a way that it will not again fill the pits. A description of the pits will be embodied in the field notes, and will include, in every instance, a statement of their size and position; this is particularly important in view of the fact that the practice herein outlined differs materially (in the interest of simplicity) from that set forth in earlier editions of the Manual.

ARRANGEMENT AND MARKING OF CORNER ACCESSORIES.

**330.** *Standard township corners.*
*Standard section corners.*

Two bearing trees, one in each section north of the standard parallel, each marked "S C" and the township, range and section; as

T 25 N R 18 E S 31 S C B T.

Mound of stone, north of corner.

Three pits, one each on line north, east and west.

**331.** *Closing township corners.*
*Closing section corners.*

Two bearing trees, one in each section to the right and left of the closing line, each marked "C C" and the township, range and section; as

T 24 N R 18 E S 6 C C B T.

Mound of stone, on the closing line.

Three pits, one on the closing line and one each to the right and left on the line closed upon.

**332.** *Corners common to four townships.*

Four bearing trees, one in each section, each marked with the township, range and section; as

T 22 N R 17 E S 1 B T.

Mound of stone, south of corner.

Four pits, one each on line north, east, south and west.

**333.** *Corners common to two townships only.*

Two bearing trees, one in each section cornering at the monument, each marked with the township, range and section; as

T 2 N R 7 W S 1 B T.

Mound of stone, on the line between the two townships cornering at the monument.

Three pits, one each on the three lines connecting at the monument.

**334.** *Corners referring to one township only.*

Two bearing trees, both in the township cornering at the monument, each marked with the township, range and section; as

T 23 N R 19 W S 36 B T.

Mound of stone, in the township cornering at the monument, at 45° from cardinal direction at the monument.

Two pits, one each on the two lines connecting at the monument.

**335.** *Corners common to four sections.*

Four bearing trees, one in each section, each marked with the township, range and section; as

T 26 N R 17 E S 35 B T.

Mound of stone, west of corner.

Four pits, one in each section northeast, southeast, southwest and northwest.

**336.** *Section corners common to two sections only.*

Two bearing trees, one in each section cornering at the monument, each marked with the township, range and section; as

T 14 S R 17 E S 12 B T.

Mound of stone, on the line between the two sections cornering at the monument.

Two pits, one in each section at 45° from cardinal direction at the monument.

**337.** *Section corners referring to one section only.*

Two bearing trees, both in the section cornering at the monument, each marked with the township, range and section; as

T 27 N R 16 W S 17 B T.

Mound of stone, in the section cornering at the monument, at 45° from cardinal direction at the monument.

Two pits, one 3 feet and one 6 feet distant, both in the section cornering at the monument, at 45° from cardinal direction at the monument.

**338.** *Standard quarter-section corners.*

Two bearing trees, both north of the standard parallel, each marked "¼" and "S C" and the section; as

¼ S 36 S C B T.

Mound of stone, north of corner.

Two pits, one each on line east and west.

**339.** *Quarter-section corners of maximum control.*

Two bearing trees, one in each section, each marked "¼" and the section; as

¼ S 16 B T.

Mound of stone: (*a*) On a meridional line, west of corner; and, (*b*) on a latitudinal line, north of corner.

Two pits, one in each direction on the line passing through the monument.

**340.** *Quarter-section corners of minimum control.*

Two bearing trees, both in the particular section which is concerned, each marked "¼" and the section; as

¼ S 7 B T.

Mound of stone, in the particular section which is concerned, in a cardinal direction from the monument.

Two pits, one in each direction on the line passing through the monument.

**341.** *Meander corners.*

Two bearing trees: (*a*) On a standard parallel or other line controlling surveys to one side only, both in the particular section which is concerned; and (*b*) on all other lines, one in each section to the right and left of the line; all marked "M C" and with the township, range and section; as

T 25 N R 14 E S 32 M C B T.

Mound of stone, on the surveyed line on the opposite side of the monument from the meanderable body of water.

Two pits, one 3 feet and one 6 feet distant, on the surveyed line on the opposite side of the monument from the meanderable body of water.

**342.** *The interior quarter-section and all sixteenth-section corners,* when required by the written special instructions.

Two bearing trees, marked (with letters and figures ending in "B T") as shown in the following diagram:

Mound of stone, in a cardinal direction from the monument, as shown (with symbol "  ") in the following diagram:

Two pits, in a cardinal direction from the monument, as shown (with symbol "□") in the following diagram:

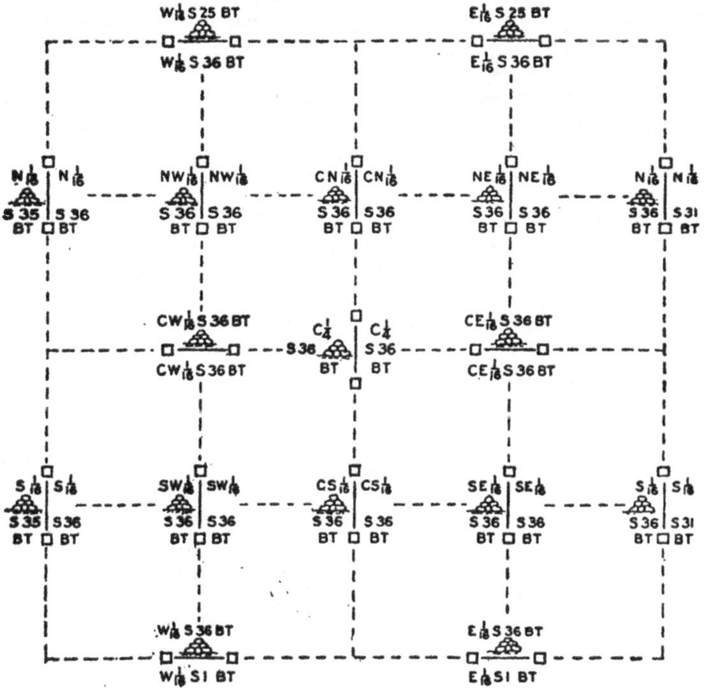

**343.** *Sixteenth-section corners of minimum control.*

Two bearing trees, both in the particular section which is concerned, each marked with a key letter (N, E, S or W) to indicate the position of the monument, and "$\frac{1}{16}$" and the section; as

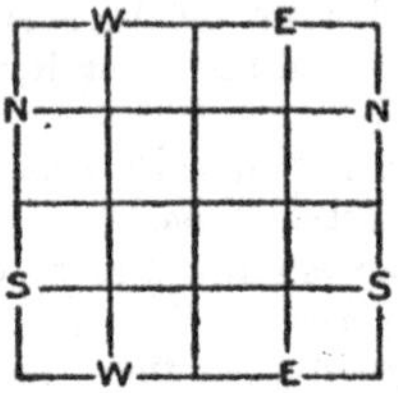


Key

N $\frac{1}{16}$ S 18 BT.

Mound of stone, in the particular section which is concerned, in a cardinal direction from the monument.

Two pits, one in each direction on the section line passing through the monument.

**344.** *Special and auxiliary meander corners.*

Two bearing trees, each marked "S M C" or "A M C," as the case may be, and the section; as

S 14 S M C B T, or
S 14 A M C B T.

Mound of stone, on the opposite side of the monument from the meanderable body of water.

Two pits, one 3 feet and one 6 feet distant, on the opposite side of the monument from the meanderable body of water.

**345.** *Closing subdivision-of-section corners.*

Two bearing trees, both in the particular section which is concerned, each marked "C C" and the section; as

S 9 C C B T.

Mound of stone, on the closing line.

Three pits, one on the closing line and one each to the right and left on the line closed upon.

**346.** *Miscellaneous angle points along irregular boundaries.*

(*a*) Two bearing trees, where the monuments are less than 1 mile apart, one on each side of the boundary; and (*b*) four bearing trees, where the monuments are 1 mile or more apart, two on each side of the boundary; each marked "A P" and a serial or section number, or both, also the initials or abbreviation of the State, reservation, grant, private claim or public land, as appropriate; as

A P 2 TR 37 B T, and<br>
A P S 14 B T (for "angle point No. 2" on the boundary of a private claim "Tract No. 37" falling on surveyed land).

Mound of stone, on the medial line between the boundary lines intersecting at the monument, and in the direction toward the State, reservation, grant or private claim.

Two pits, one in each direction on the lines intersecting at the monument.

**347.** *Intermediate corners along irregular boundaries.*

(*a*) Two bearing trees, where the monuments are less than 1 mile apart, one on each side of the boundary; and (*b*) four bearing trees, where the monuments are 1 mile or more apart, two on each side of the boundary; each marked with the number of the mile or half-mile corner and the letter "M" (to indicate "mile corner"), and the initials or abbreviation of the State, reservation, grant, private claim or public land, as appropriate; as

47 M COLO BT, and<br>
47 M OKLA BT (for "47th mile" corner on the boundary line between the States of "Colorado" and "Oklahoma").

Mound of stone, on a line at right angles to the boundary, and in direction toward the State, reservation, grant or private claim.

Two pits, one in each direction on the boundary.

# CHAPTER V.

## RESTORATION OF LOST CORNERS.

### IDENTIFICATION OF EXISTENT CORNERS.

**348.** It is the purpose of this chapter of the Manual to outline the guiding principles which are to be observed in the identification of existent corners, and thereafter to set forth the particular rules which are to be applied in the recovery of the position of lost corners originally established in the execution of the United States rectangular surveys.

All surveyors, whether employed by the United States or not, are cautioned to note the difference between the regulations pertaining to the establishment of the original surveys of the public lands and those relating to the subsequent identification of said official surveys and the replacement of missing monuments thereof.

In the extension of the rectangular surveys it devolves upon the United States surveyor to identify the initial lines of his group and to replace all lost corners thereof. On the other hand in the subdivision of sections and in the location of property lines generally, it falls to the county or other local surveyor to identify the official corners, and where a required corner is missing the local surveyor will be called upon to recover the point. Thus it will be seen that local as well as United States surveyors are constantly called upon to search for existing evidence of original monuments, and in this work the surveyors will be guided by the same general methods. Should the search for a monument result in failure, the appropriate restorative surveying process to be observed by either surveyor will be based upon the same rules as hereinafter outlined. The text that follows draws no distinction between the duties of the two classes of surveyors.

**349.** The terms "corner" and "monument" are used largely in the same sense, though a distinction should be noted to clarify the subject matter of this chapter. The term "corner" is employed to denote a point determined by the surveying process, whereas the "monument" is the physical structure erected for the purpose of marking the corner point upon the earth's surface.

**350.** An existent corner is one whose position can be identified by comparing the evidence of the monument or its accessories on the ground, with the record contained in the field notes of the original survey, or where the point can be determined otherwise by suitable testimony.

**351.** The process of again bringing to light the physical evidence of an original monument is founded on the principle of intelligent search for the calls of the field notes of the original survey, guided by the controlling influence of known points. The problems incident to the search are vastly simplified whenever a retracement may be projected from known points, and the final search for a monument should cover the zone surrounding one, two, three or four temporary points as may be determined by connections with known corners in one, two, three or four directions, according to the number of points which will ultimately control the relocation in case the corner in question should be declared lost.

**352.** The character of the original monument is the most important factor in regard to its lasting qualities, and the search should be directed to an examination for such evidence as may reasonably be expected to remain. The evidence is bound to range from that which is least conclusive to that which is unquestionable, and the requisite support of corroborative evidence is necessary in direct proportion to the uncertainty of any feature regarding whose authenticity there may be danger of dispute.

A stone, wooden post, tree corner, deposit corner, and the modern iron post monument are all subject to more or less deteriorating changes through various influences, depending upon the character of the original monument, its local site conditions, and the lapse of time, and all such factors should be taken into consideration when comparing the particular evidence in question with the description contained in the original field notes.

**353.** If the evidence of the monument is not fully conclusive, the surveyor's attention will be directed at once to the record accessories; this step is so generally necessary that it should be considered simultaneously with the search for the monument; in fact, in their broader significance the accessories are a part of the monument.

The underlying principles relating to the identification of the corner accessories, subject to the changes which may be expected in the period intervening after the date of the original survey, have already been fully outlined in Chapter IV. It will suffice to state that the evidence of the accessories should agree with the record contained in the field notes of the original survey, subject only to such changes as may reasonably be expected.

**354.** In case of material disagreement between the particular evidence in question and the record calls, the process of elimination of those features regarding which there may be doubt, after making due allowance for natural changes, will serve a most useful purpose, as follows:

(*a*) The character and dimensions of the monument in evidence should not be widely different from the record;

(*b*) The markings in evidence should not be inconsistent with the record; and,

(*c*) The nature of the accessories in evidence, including size, position and markings, should not be greatly at variance with the record.

A certain measure of allowance for ordinary discrepancies should enter into the consideration of the evidence of a monument and its accessories, and no definite rule can be laid down as to what shall be sufficient evidence in such cases. Much must be left to the skill, fidelity and good judgment of the surveyor in the performance of his work, ever bearing in mind the relation of one monument to another, and the relation of all to the recorded natural objects and items of topography.

**355.** A corner will not be considered as lost if its position can be recovered satisfactorily by means of the testimony and acts of witnesses having positive knowledge of the precise location of the original monument. The expert testimony of surveyors who may have identified the original monument prior to its destruction and thereupon recorded new accessories or connections, etc., is by far the most reliable, though landowners are often able to furnish valuable testimony. The greatest care is necessary in order to establish the bona fide character of the record intervening after the destruction of an original monument. Full inquiry may often serve to bring to light various records relating to the original corners, and memoranda of private markings, etc., and the surveyor should make use of all such sources of information. The matter of boundary disputes should be carefully looked into in so far as adverse claimants may base their contentions upon evidence of the original survey, and if such disputes have resulted in a boundary suit, the record testimony and the court's decision should be carefully examined relative to any information which may shed light upon the position of an original monument.

The testimony of individuals may relate to knowledge of the original monument or the accessories, prior to their destruction, or to any other marks fixing the locus of the original survey, and the value of such testimony may be weighted in proportion to its completeness and agreement with the calls of the field notes of the original survey, also upon the steps taken to preserve the location of the original marks. All such evidence should be put to the severest possible tests by confirmation relating to known original corners and other calls of the original field notes, particularly to line trees, blazed lines and items of topography.

It is impossible to outline a definite rule for the acceptance or non-acceptance of the testimony of individuals. Corroborative evidence becomes necessary in direct proportion to the uncertainty of the particular statements advanced by the individual who testifies. It will be well for the surveyor to bear in mind that conflicting statements and contrary views of interested parties are fruitful of boundary disputes.

**356.** In those cases where witness corners were established in the original survey, the true point for the corner will be controlled by such witness corner, when the latter can be identified, by reference to the record in accordance with the general plan of the survey. The usual diligent search will be made for witness corners, but where the same can not be identified the position of the true point for the corner will usually be of major importance, rather than the point for the witness corner, and in such instances the surveyor will proceed directly to the re-determination of the true corner position, adopting the particular methods which should govern the case in hand. Should it become necessary to restore a lost witness corner the general principles hereinafter outlined will be observed.

**357.** In the absence of an original monument, a line tree, or a definite connection to natural objects, or to improvements, which can be identified, may each fix a point of the original survey for both latitude and departure. The mean position of a blazed line, when identified as the original line, may sometimes help to fix a meridional line for departure, or a latitudinal line for latitude. Other alls of the original field notes in relation to various items of topography may assist materially in the recovery of the locus of the original survey. Such evidence may be developed in an infinite variety.

It may be only such as to disprove other questionable features, or it may guide the surveyor in a general way in arriving at the immediate vicinity of a line or corner, or in its best phases may be such as to fix the position of a line or corner beyond any doubt.

**358.** A certain measure of allowance should be made for ordinary discrepancies in the calls relating to items of topography. Such evidences should be considered more particularly in the aggregate, and when they are found to be corroborative an average may be secured to control the final adjustment, which will be governed largely by the evidences nearest the particular corner in question, giving the greatest weight to those features which agree most harmoniously with the record, and to such items as afford definite connection. A careful analysis will generally reveal the merits of authentic evidences as opposed to unreliable features bearing resemblance to the calls of the field notes, and in this matter the surveyor will find an opportunity to exercise his skill to the fullest capacity.

**359.** It is a matter of utmost importance to determine where an identified call of the original field notes shall operate to control for both latitude and departure, or for either coördinate by itself, and finally as to the necessity for applying the rules for proportionate measurement where the distance between the identified points is considerable.

### RESTORATION OF LOST CORNERS.

**360.** A lost corner is a point of a survey whose position can not be determined, beyond reasonable doubt, either from original traces or from other reliable evidence relating to the position of the original monument, and whose restoration on the earth's surface can be accomplished only by means of a suitable surveying process with reference to inter-dependent existent corners.

**361.** The surveyor is not prepared to consider the restoration of a lost corner until he has exhausted every other means of identifying its original position, and at this stage of his work he should have determined upon an approximate position of the original monument based upon his findings resulting from retracements leading from known corners to the lost corner, from one, two, three or four directions in accordance with the plan of the original survey. The principle of proportionate measurement, which most nearly harmonizes surveying practice with the legal and equitable considerations involved in controversies concerning lost land boundaries, enters into the problem at this stage, and this plan of relocating a lost corner will always be employed unless outweighed to the contrary by physical evidence of the original survey. In cases where the relocated corner can not be made to harmonize with all the calls of the original field notes, due to unexplained discrepancy which is made apparent by the retracement, the surveyor is required to determine which calls will be given major control, and those which must be subordinated.

**362.** The preliminary retracements furnish the only possible means of arriving at the discrepancies of the courses and distances of the original survey as compared with those derived in the process of re-running the lines, and the whole problem of proportionate measurement is one involving the adjustment of said discrepancies. The restoration of the lost corners can not proceed until the retracement of the original survey has been completed. The retracement will be based upon the courses and distances returned in the field notes of the original survey, or the equivalent by calculation, initiated and closed upon known original corners. Temporary stakes for future use in the relocation of all lost corners may be set when making the retracements.

**363.** As has been observed, existing original corners can not be disturbed; consequently discrepancies between the new and the original record measurements of the line connecting the identified original corners will not in any manner affect measurements beyond said corners, but the differences will be distributed proportionally within the several intervals embraced in the line in question.

**364.** A proportionate measurement is one resulting in concordant relation between all parts of an original record length of a line and the new distances given to the several parts as determined by the re-measurement, in such a manner that the new distance given to any part of a line shall bear the same relation to the original record length of that part of the line as the new measurement of the whole line bears to the original record length of said line. The ordinary field problem consists in distributing the excess or deficiency determined by comparing the new measurement with the record distance between two original existent monuments, in such a manner that the amount of excess or deficiency given to each interval shall bear the same proportion to the whole difference as the record length of the interval bears to the whole record distance. After having applied the proportionate difference to the record length of each interval the sum of the adjusted lengths will equal the new measurement of the whole distance.

365. The term "single proportionate measurement" is applied to a new measurement made on a single line to determine the position thereon for restoring a lost corner, for example, a quarter-section corner on line between two original section corners. The term "double proportionate measurement" is employed to signify new measurements made between four original corners on intersecting meridional and latitudinal lines for the purpose of fixing by relation to both lines the position of a lost corner, for example, a corner common to four sections or four townships.

366. It will almost invariably happen that discrepancies will be developed between the new measurements and the original measurements recorded in the field notes. When these differences occur the surveyor will generally be required to adopt a proportionate measurement based upon a process conforming to the method followed in the original survey. The principle of the preponderance of one line over another of less importance is recognized, in order to determine upon the procedure relative to single or double proportionate measurement, or other rule to be adopted in order to limit the control and at the same time harmonize the restorative process with the method followed in the original survey. Thus standard parallels will be given precedence over other township exteriors, and the latter will be given precedence over subdivisional lines; section corners will be relocated before the position of lost quarter-section corners can be determined.

## PRIMARY METHODS.

### (a) DOUBLE PROPORTIONATE MEASUREMENT.

367. The method of double proportionate measurement is generally applicable to the restoration of lost corners of four townships and of lost interior corners of four sections. It is the best example of the basic principle that monuments north and south should control the latitudinal position of a lost corner, and monuments east and west should control the longitudinal position of a lost corner, upon a plan by which the influence of one identified original corner is balanced by the control of a corresponding original corner upon the opposite side of a particular missing corner which is to be restored, each identified original corner being given a controlling weight inversely proportional to its distance from the lost corner.

55465°—19——18

**368.** In order to restore a lost corner of four townships where all of the connecting lines have been established in the field, a retracement will first be made between the nearest identified original corners on the meridional line, north and south of the missing corner, upon which line a temporary stake will be placed at the proper proportionate distance. This will determine the latitude of the lost corner. Next, the nearest original corners on the latitudinal line will be connected and a point thereon will be determined by proportionate measurement in a similar manner, independent of the temporary stake on the meridional line. The second temporary point will determine the position of the lost corner in departure. Then through the first temporary stake run a line east or west, and through the second temporary stake a line north or south, as relative situations may determine. The intersection of the two lines last run will define the position of the restored corner by "double proportionate measurement."

**369.** In the accompanying diagram the points "A," "B," "C" and "D" (on the small scale) represent four original corners; and (on the large scale) "E" represents the proportional point between "A" and "B," for measurement only, and similarly, "F" represents the proportional point between "C" and "D." The point "X" satisfies the first control for latitude, and the second control for departure.

**370.** The plan of double proportionate measurement will be applied to the restoration of lost corners of four townships where all the lines therefrom have been run. Lost interior corners of four sections, where all the lines therefrom have been run, will also be reestablished by double proportionate measurement, *after* first relocating the required lost section corners on the township exteriors. When a number of corners of four sections, and the intermediate quarter-section corners, are missing on all sides of the one sought to be reestablished, the entire distance must, of course, be remeasured between the nearest identified corners both north and south, and east and west, in accordance with the rule laid down.

**371.** Where one of the connecting lines has not been established in one direction from the missing township or section corner, the record distance to the nearest identified corner in the opposite direction will prevail in lieu of a proportional measurement. Thus, in the same diagram, if the latitudinal line in the direction of the point "D" had not been established in the original survey, the

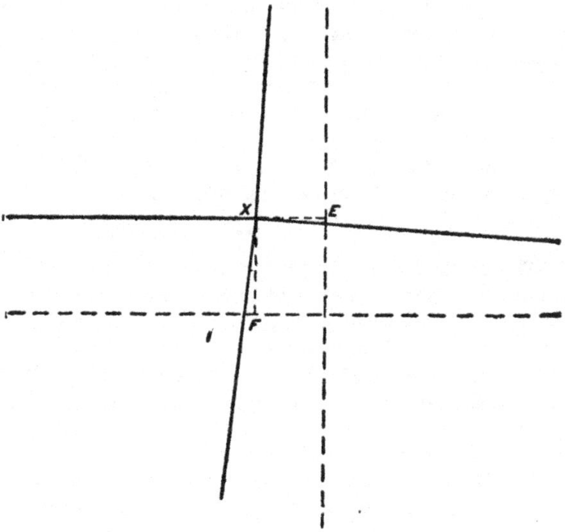


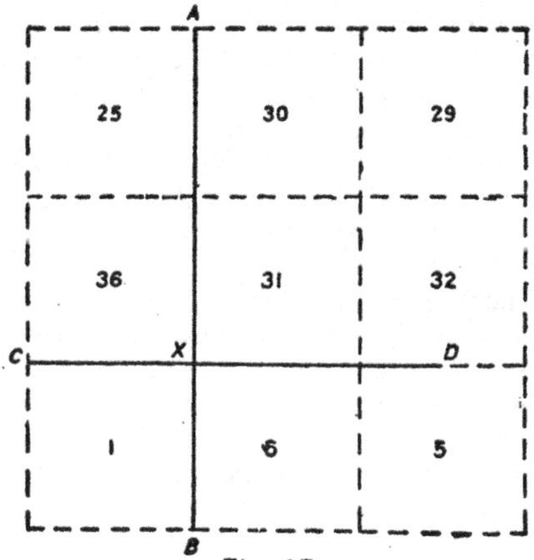


Fig. 67.

position of the point "F" in departure would have been determined by reference to the record distance from the point "C," whereupon the point "X" would have been fixed by cardinal offsets from the points "E" and "F" as before. Again, in rare instances, where the intersecting lines have been originally established in only two of the directions, the record distances to the nearest identified corners on the two lines will control the position of the temporary points from which the cardinal offsets are to be made.

(b) SINGLE PROPORTIONATE MEASUREMENT.

**372.** The method of single proportionate measurement is generally applicable to the restoration of lost corners on standard parallels and other lines established with reference to definite alinement in one direction only. Intermediate corners on township exteriors and other controlling boundary lines are to be included in this class.

**373.** In order to restore a lost corner by single proportionate measurement, a retracement will be made connecting the nearest identified regular corners upon the particular line in question, the record of which shows no deflection in alinement; a temporary stake will be set on the preliminary line at the original record distance; the total distance will be measured, also the falling at the objective corner. The temporary stake will then be adjusted for the proportional part of the difference between the record distance and the re-measurement, also for its proportional part of the falling. Thus the adjusted position will fall on the true line connecting the nearest identified corners, and at the same proportional interval from either as existed in the original survey. Any number of lost points, on the same straight line, may be recovered by the same plan, setting a temporary corner for each at the time when making the retracement. On the retracement of an east and west line, the proper adjustments to secure the true latitudinal curve should be allowed for as outlined in Chapter II.

**374.** Lost standard corners will be restored to their original positions on a base line, standard parallel or correction line, by single proportionate measurement on the line connecting the nearest identified original standard corners on opposite sides of the missing corner or corners, as the case may be. The term "original standard corners" will be understood to designate standard township, section and quarter-section corners, meander corners terminating the survey of a standard parallel, and closing corners in those cases where they were originally established during the survey of a standard parallel as corners from which to project surveys to the south. No other meander or closing corners along a standard parallel will control the restoration of lost standard corners.

**375.** All lost exterior section and quarter-section corners will be restored by single proportionate measurement between the nearest identified corners on opposite sides of the missing corner, north and south on a meridional line, or east and west on a latitudinal line, *after* the township corners have been identified or relocated. An exception to this rule will be noted in the case of any exterior the record of which shows irregularities in alinement between the terminal township corners. (See sec. 380.)

**376.** All lost interior quarter-section corners will be restored by single proportionate measurement between the adjoining section corners, *after* the section corners have been identified or relocated.

**377.** Lost meander corners, originally established on a line projected across the meanderable body of water and marked upon the opposite side thereof will be relocated by single proportionate measurement, *after* the section or quarter-section corners upon the opposite sides of the missing meander corner have been duly identified or relocated.

(c) CLOSING CORNERS.

**378.** In order to reestablish a lost closing corner on a standard parallel or other controlling boundary, the line closed upon will be retraced, beginning at the corner on the standard parallel or other controlling boundary from which the connecting measurement was originally made, itself properly identified or relocated; a temporary stake will be set at the original record connecting distance, and the total distance and falling will be noted at the next regular corner on the opposite side of the missing closing corner. The temporary stake will then be adjusted as in single proportionate measurement, i. e., the closing corner will be reestablished on the true line closed upon at the proper proportional interval between the nearest regular corners to the right and left. An identified closing corner not actually located in the line closed upon will determine the direction of the closing line, but not its legal terminus; the latter is bound to fall at the true point of intersection of the two lines. The position of a restored closing corner should be verified by a retracement of the line whose terminus it was designed to mark. (See sec. 384.)

SECONDARY METHODS.

379. The following methods involve special applications of the general rules of proportionate measurement for adoption in unusual cases where the ordinary control can not be obtained.

(d) BROKEN BOUNDARIES.

380. In order to restore one or more lost corners on a broken or irregular township exterior, or other controlling boundary, a retrace-

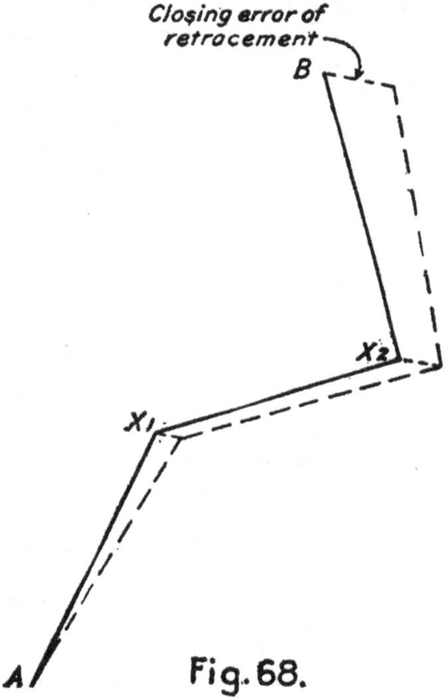


Fig. 68.

ment will be initiated at the nearest identified original corner on the boundary, following out the record courses and distances, or the equivalent by calculation, setting a temporary stake for each missing corner or angle point, until the next identified original corner has been attained, where a final temporary stake will be set at the record distance of the last course of the retracement. The closing error will then be determined for course and distance from the last temporary stake to the objective original corner, and each temporary stake will thereafter be adjusted on the bearing of the closing error, a proportional amount of the length of the closing error equal to the proportional part of the distance of the temporary stake from the initial point of the retracement, i. e., the particular distance to be measured at any temporary stake, on the bearing of the closing error, is to the whole length of the closing error as the distance of the particular temporary stake from the initial original corner is to the whole length of the retracement. Angle points and intermediate corners will be treated alike.

(e) ORIGINAL CONTROL.

**381.** Where a line has been terminated with reference to a measurement in one direction only, a lost corner will be restored by reference to the original record bearing and distance, counting from the nearest regular corner, the latter having been duly identified or restored. Examples will be found where lines have been discontinued at the intersection with large meanderable bodies of water, or at the border of what was classed as impassable ground.

(f) INDEX CORRECTION FOR AVERAGE ERROR IN ALINEMENT AND MEASUREMENT.

**382.** In unusual cases where a retracement has been made of many miles of the original lines, between identified original corners, and there has been developed a definite surplus or deficiency in measurement, or a definite variation in alinement, characterizing the original survey, it will be proper to make allowance for such average "index error." Such adjustment will be taken care of automatically in all cases where there exists a suitable basis for proportional measurement, but in any case where such control is lacking, an index error, if conclusive, will be made use of by applying the determined correction to the record courses and distances. If there is not conclusive evidence of such index error the record courses and distances will be allowed to prevail.

SPECIAL CASES.

**383.** Examples of special cases could be set forth almost indefinitely, but without bringing out important new principles. In some respects the treatment of a large number of special examples would serve to confuse the subject by seeming to warrant certain procedure as a general rule which in fact would not be proper were the conditions altered; the latter occur in an infinite variety. Ample provision has been made for the United States surveyor to call upon a supervising officer for advice in difficult cases, and where necessary the latter is in a position to direct the surveyor to proceed with additional retracements in order to develop any data which should be considered before a decision is rendered. In trials of boundary suits the court will generally consider many additioual questions besides those concerned in the technical problem, and in such instances an academic study of hypothetical examples might serve to cloud the real issue. It would be beyond the purpose of the Manual to invade the realm of non-technical matter while attempting to lay down the general principles involved in the restoration of lost corners.

**384.** In all unusual instances, where on account of manifest distortion, or through extensive obliteration resulting in great distances between existing corners, or otherwise, the evidence of a survey can not be identified with sufficient certainty to enable a suitable application of the various rules relating to the restoration of lost corners, the surveyor is again advised to report the facts to the proper supervising officer. In the same connection, it is important that the surveyor should not be confused with the notion that he is required, or has any authority, to revert to the principles relating to the establishment of original surveys as an alternative in such cases. The methods incident to resurveys, as outlined in the next chapter, are designed to rectify unusual conditions which are widely at variance with the representations of the original approved plat and field notes.

### (*g*) MISCELLANEOUS CONTROL.

**385.** It will be apparent to the experienced surveyor that actual field conditions do not always furnish the basis for the application of the rules heretofore set forth, and while developing a consistent theory to apply in unusual cases the surveyor will at once note that the first consideration relates to a more or less arbitrary limitation of the control to be adopted. No definite rule can be laid down, except that there should be the closest possible adherence to the basic examples already given in the text. The methods heretofore outlined readily harmonize surveying practice with legal decisions concerning the restoration of lost corners. A strictly consistent mathematical recovery of a lost corner, not based upon any known legal decision, may be obtained by allowing every known corner within a reasonable radius to enter into the control, each original corner being given a weight inversely proportional to its distance from the missing corner, and though the principle will lead to the same result in come cases as by the methods previously outlined, it will yield a slightly different result under other regular circumstances. For the latter reason a miscellaneous control based upon such mathematical principle will not be adopted except as specifically approved by the proper supervising officer after due consideration of the facts in regard to the applicability of the method in the absence of a suitable basis for a regular control.

**386.** Having thus safeguarded the application of the following method, the problem in the field will be developed by a series of retracements each beginning at an accepted corner, thence following out the record courses and distances, each retracement terminating at a temporary stake in the vicinity of the objective lost corner. Each stake will be given a weight inversely proportional to the distance from the accepted corner to which it is related. The several temporary stakes will then be combined; the first two to be resolved into a point on the line between them, dividing the whole distance into two parts that will make the interval from either stake inversely proportional to the weights previously assigned, and the latter point will be given their combined weights. The last point will then be correlated with the third temporary stake on a similar plan. Three or more original corners will thus exercise their influence upon the final resultant position for the corner which is to be restored. The result will be the same no matter what the order of connecting the temporary stakes may be, but the omission of any element of the control or the introduction of an additional original corner will alter the final position. The field of influence should accordingly be selected with a view to obtaining a resultant balanced position which can not be materially changed by the introduction of other known points of control.

# CHAPTER VI.

## RESURVEYS.

### JURISDICTION.

**887.** Certain important considerations are involved in the execution of Government resurveys of an entirely different character from those relating strictly to the making of original surveys; these considerations present matters not referred to in Chapter V. There is a twofold object of a resurvey: First, the adequate protection of existing rights acquired under the original survey in the matter of their location on the earth's surface, and, second, the proper marking of the boundaries of the remaining public lands.

**888.** As already noted in Chapter I, the Congress has authorized, under certain conditions, the re-marking of the public-land surveys. The acts relating to resurveys contemplate a restoration of the corners of the original surveys in those townships, (*a*) where the obliteration of the original monuments or other evidence of the position of the original lines has become so advanced that the land boundaries can be identified only through extensive retracements by experienced surveyors of the General Land Office, and (*b*) where field investigation shows that conditions on the ground disagree with the representations upon the original plat to such an extent that the land boundaries can not be identified positively in one position to the exclusion of another, in consequence of which said plat should be disqualified as a basis for the disposal of remaining public land. While the Government may initiate a resurvey in the absence of any application therefor, as a rule, the steps preliminary to the authorization of a resurvey will be taken by the settlers interested in the land, through a showing of facts made to the proper supervising officer, setting forth the existing conditions with respect to the original survey and status of ownership of the lands.[1]

---

[1] See current circular governing applications for resurveys.

**389.** The surveyor is advised to bear in mind the fact that in localities where resurveys are necessary the occasion for boundary disputes is ever present; he should accordingly exercise the greatest care in his technical work in the field and in the record thereof, so that the result of the resurvey shall relieve existing difficulties as far as possible without introducing new complications. As in the case of original surveys, the records of all resurveys must form an enduring basis upon which depends the security of the title to all lands acquired thereunder, and the field notes should be so prepared that under the test of the closest possible scrutiny at all times, present and future, the record can be regarded as conclusive in the matter of the location of such rights.

**390.** The General Land Office has exclusive jurisdiction over all matters pertaining to surveys and resurveys affecting the public lands; as between private owners of lands the title to which has passed out of the United States, final determination in the matter of fixing the position of disputed land boundaries rests with the local court of competent jurisdiction. The rules of procedure laid down by the General Land Office to guide its surveyors in the re-marking of lines of previous surveys are intended to be in harmony with the leading court decisions in suits involving boundary disputes, and said rules should be so applied that the courts may, with security, accept without question the boundaries thus determined in so far as they represent the true location of a particular tract intended to be conveyed by a patent. Government resurveys are undertaken only by duly appointed United States surveyors acting under the authority of the Secretary of the Interior through the Commissioner of the General Land Office and under the immediate direction of subordinate supervising officers.

### LIMIT OF AUTHORITY OF SURVEYOR.

**391.** There are certain questions of a purely judicial nature involved in resurveys of every description where the decision is to be reserved to the General Land Office, particularly those relating to compliance with the general laws in respect to the entry of the public lands. Thus it comes within the realm of the surveying process to identify and mark out on the ground the various legal subdivisions of the public domain, but it is a judicial question beyond the function of the surveyor to determine whether or not specified lands have been duly earned under a certain entry. In the resurvey process the surveyor will determine whether or not lands embraced within a claim as occupied have been correctly related in position to the original survey, and where the demonstration of this question may be one involving more or less uncertainty, as is often the case, the surveyor will examine and weigh the evidence relating strictly to the surveying problem involved, and he will interpret the evidence in respect to its effect upon the manner in which the resurvey shall be executed looking to the protection of the valid rights acquired under the original survey. The surveyor has no authority to enter into any agreements looking to the exchange of one subdivision for another, or to bind the General Land Office in this particular.

### BONA FIDE RIGHTS OF CLAIMANTS.

**892.** In order to carry out the provisions of the laws relating to resurveys, the surveyor should understand fully the meaning of the words "bona fide rights" and under what circumstances it will be held that such rights have been impaired by a resurvey. In this connection attention is again directed to the clause contained in the act of March 3, 1909 (35 Stat., 845), as amended by joint resolution approved June 25, 1910 (36 Stat., 884), which reads as follows:

> "That no such resurvey or retracement shall be so executed as to impair the bona fide rights or claims of any claimant, entryman, or owner of lands affected by such resurvey or retracement."

The rights of claimants are to be given similar protection under the provisions of the act of September 21, 1918 (40 Stat., 965).

**893.** It will be understood that bona fide rights are those acquired in good faith under the law. Rights of this character can be affected by a resurvey only in the matter of POSITION OF LOCATION on the earth's surface, and the surveyor will be concerned only with the question as to whether lands covered by such rights have been actually LOCATED in good faith. Other questions of good faith, such as priority of occupation, possession, continuous residence, value of improvements, and cultivation, when considered apart from the question of the position of the original survey, do not in any manner affect the problem of resurvey.

It is evident that the resurvey must afford adequate protection to bona fide rights vested in both improved and unimproved lands. In the final determination of the true position of all lands, whether improved or unimproved, in the absence of original corners, the necessity for more or less flexibility of method must be recognized, as the value of both of these classes of lands *may* be vitally affected by an arbitrary process of resurvey which is rigid in its application. Unimproved lands, however, where no apparent attempt has been made on the part of the owner to identify the same under their original descriptions (and where the inherent value of the lands in question is the same), are not *necessarily* affected in the same manner, and such unimproved lands may be adjusted to a position found by the surveyor to be conformable to adjoining or near-by tracts, where all may be held to qualify under the rule of acceptable location.

**394.** The question arises whether the technical rules for the restoration of lost corners are to be rigidly applied in all cases regardless of their effect on the position of improvements, or whether the position of all improvements is to be accepted without question regardless of the relation or irrelation of such improvements to the existing evidence of the original survey and to the description contained in the entry. Manifestly these opposite extremes are equally unacceptable. Somewhere between them, therefore, will be found the basis for a determination of the question as to when lands so improved are to be regarded as having been LOCATED in good faith or otherwise. It is clear that no definite specific set of rules can be laid down in advance for the determination of this question. This is a problem the solution of which must be found on the ground by the surveyor; it is upon his judgment primarily that the responsibility for a determination of the question of good faith as to LOCATION must rest. The surveyor may err in his judgment, but once this question is settled to his own satisfaction, the *procedure* to be adopted in the matter of the application of resurvey rules is no longer in doubt.

**395.** It may be held generally that an entryman has located his lands in good faith (referred to herein as an acceptable location of a claim or of a local point), when it is evident that his interpretation of the record of the original survey as related to the nearest existing corners at the time the lands were located (as defined by his fencing, culture, or other improvements) is indicative of such a degree of care and diligence upon his part, or that of his surveyor, in the ascertainment of his boundaries, as might be expected in the exercise of ordinary intelligence under existing conditions. From this it follows that lack of good faith is not necessarily chargeable against an entryman if he has not located himself according to a rigid application of the rules laid down for the restoration of lost corners, where complicated conditions involve a double set of corners, both of which may be regarded as authentic; or where the nearest existing corners in one or more directions are an excessive distance away; or are improperly related to each other to an extraordinary degree; or where all evidences of the original survey which had been adopted by the entryman as a basis for his location have been lost before the resurvey is undertaken. Furthermore, the extent of recognition given by neighboring claimants to a local point used for the control of the location of claims very often carries with it the necessity for a consideration by the surveyor of its influence in the matter of the acceptability of such locations under the foregoing rule of good faith.

**396.** In cases involving extensive obliteration at the date of entry, the entryman or his successors in interest may be charged with the knowledge that the boundaries of the claim will probably be subject to more or less adjustment in the event of a resurvey, and that in the process of fixing the boundaries of groups of claims a general control applied to all must be favored as far as possible in the interest of equal fairness to all and of simplicity of resurvey. Even in the presence of extensive obliteration of the original survey, a claim which manifestly shows that no attempt has been made to relate the same in some manner to the original survey can not generally be regarded as having been located in good faith.

**397.** Cases will arise where it may be evident that lands have been OCCUPIED in good faith, but whose boundaries as occupied are clearly in disagreement with the demonstrated position of the legal subdivisions called for in the description. Obviously the rule of good faith as to location can not apply, and relief must be sought through the process of amended entry (act of Feb. 24, 1909, 35 Stat., 645) to cover the legal subdivisions actually earned, rather than through an alteration of the position of established lines. This is a process of adjudication rather than one of resurvey. A case of this character should be regarded as an "erroneous location," in precisely the same manner as would obtain if the question of resurvey were not involved.

**398.** The recognition of the principle that the restoration of a corner may be influenced by the position of one or more existing claims warrants, within suitable limits, the acceptance of an unofficial determination, in the manner hereinafter stated, which would not necessarily agree with that resulting from a rigid application of arbitrary rules laid down for the restoration of lost corners.

### GENERAL FIELD METHODS.

**399.** There are two recognized methods of making Government resurveys—DEPENDENT and INDEPENDENT—and in general, any field condition that may arise can be taken care of by the application of one or the other method.

**400.** The DEPENDENT resurvey is designed to accomplish a restoration of what purports to be the original conditions according to the record, based, first, upon identified existing corners of the original survey and other recognized and acceptable points of control, and, second, upon the restoration of missing corners by proportionate measurement in harmony with the record of the original survey. This type of resurvey is applicable to those cases showing fairly concordant relation between conditions on the ground and the record of the original survey. Titles, areas and descriptions should remain absolutely unchanged in the typical dependent resurvey.

**401.** The INDEPENDENT resurvey provides methods adapted to considerable areas of public land where the original survey can not be identified with any degree of certainty in accordance with the representations of the approved plat and field notes, and where the prevailing conditions are such that strictly restorative processes, when applied as an inflexible rule between existing monuments or adopted corner positions, are either inadequate or lead to unsatisfactory results. This type of resurvey provides for the segregation of individual tracts when necessary, or a conformation of individual tracts to the subdivisions of the resurvey if suitable. These processes are found to be more flexible in their application than those of the strictly dependent type, but at the same time they duly protect all private rights which have been acquired upon the basis of the original approved survey and plat. With respect to the identification and description of the public lands involved, the independent type of resurvey supersedes the record of the original survey. This will be made apparent by the representations of the approved resurvey plat.

**402.** The basic principle, with respect to the protection of bona fide rights, involved in one type of resurvey is identical with that of the other type, whether dependent or independent; they are both to be regarded as a demonstration, on the part of the General Land Office, in the light of the *best evidence available*, by means of the legal subdivisions of a dependent resurvey or by the tract segregations of an independent resurvey, of the *original* position of entered or patented legal subdivisions or lots included in the *original* description when related to the *original* survey.

**403.** The necessity for both types of resurvey is encountered in the field; the applicability of one or the other method is altogether a question depending upon local conditions, such as extent of obliteration, relative harmony of identified and recognized points, and extent of disposals by the Government. These questions should not be judged in advance of a comprehensive field examination.

**404.** In general, a preliminary field examination will be required and authorized before the resurvey is to be undertaken.

The purpose of an investigation is to develop the extent of the obliteration of the evidence of the original survey, the extent of settlement, the agricultural possibilities of the township, and any other information from which the necessity for, and the propriety of, the proposed resurvey may be determined.

A second purpose to be subserved by an investigation is the assembling of sufficient data concerning the local survey conditions to permit a proper type selection; and with this end in view the examining surveyor should investigate and report upon the relative position of the evidence of the original survey; the degree to which identified points are concordant or the reverse; the extent to which corners discordantly related have been made the basis of claim locations; the presence of one or more systems of unofficial local surveys which have been recognized and adopted by the claimants in fixing their boundaries; and the degree to which conflicts are to be anticipated.

**405.** The proper supervising officer will provide the examining surveyor with suitable instructions in which the scope of the examination will be indicated and attention will be directed to the particular considerations which should receive attention. During the progress of the investigation interested parties should be informed, upon inquiry, that the work then in progress is merely preliminary and only for the purpose of gaining information, and that if resurvey is ultimately authorized all valid rights will then be protected as required by law.

55465°—19——19

**406.** The examiner's report should contain definite recommendations concerning the type of resurvey which, in his judgment, should properly be applied in view of the prevailing conditions.

When the report and recommendations of the examiner, with those of the supervising officer, have been received by the General Land Office, the situation will be considered, the appropriate type of resurvey will be determined, and the preparation of special instructions for the resurvey will be authorized.

**407.** The special instructions, which must of necessity be based largely upon the data provided by the examination, will indicate the scope of the work, and, regardless of whether the lands are to be dependently or independently resurveyed, the necessary retracements will be made to fix the outboundaries of the township or townships designated for resurvey. With the limiting boundaries once restored so as to protect under the rules already laid down all existing property rights in the adjoining lands not to be resurveyed, the plan of procedure outlined in the instructions should, under the known conditions, produce satisfactory results, and adherence thereto is expected. If, however, unforeseen conditions are developed in the progress of the resurvey, which may apparently render the special instructions inapplicable or likely to produce inconsistent or unsatisfactory results, it is of the utmost importance that the surveyor suspend further monumentation of the corners; and after such additional retracement and investigation as may be necessary to a proper understanding of the situation, he should report the facts to the proper supervising officer and request further instructions.

**408.** During the progress of the resurvey the surveyor should advise all interested parties, as occasion and opportunity may offer, that the resurvey is not to be regarded as official or binding upon the United States until duly accepted by the Commissioner of the General Land Office, as provided by law, and that no contemplated alteration in the position of improvements or claim boundaries should be made in advance of the official acceptance of the resurvey.

## THE DEPENDENT RESURVEY.

### GENERAL CONTROL.

**409.** A dependent resurvey is an official re-marking of the original lines upon a plan whereby existing evidence of the original survey is given primary control over the position of the lines to be reestablished. A certain amount of flexibility (as hereinafter described) is allowable in the dependent resurvey when necessary for the protection of bona fide rights of claimants, particularly in those cases where no objection is found to adopting a point acceptably located under the rule of good faith already laid down, when only slightly at variance with the theoretical position of the same.

**410.** In theory the process consists, first, in the retracement and reestablishment of the township exteriors; second, the identification of all existing interior corners or other evidence of the original survey; and, third, the determination, by a *suitable field procedure*, of the theoretical position of all missing corners as indicated by a proper interpretation of the record of the original survey in relation to such existing evidence. The actual field process may be varied to some extent in order to meet local conditions or to suit the convenience of the surveyor, but the theoretical position finally determined must be identical with that which would result from a strict application of the principles of proportional measurement. When this has been accomplished, attention should be given to the adoption, as an integral part of the resurvey system, of corner positions determined by the evidences, of whatever character, of acceptable claim location. Such evidences may, for convenience, be termed "collateral evidence" as distinguished from direct evidence of the original survey.

**411.** The process of the dependent resurvey differs in scope from that applied for the usual restoration of one or more lost corners, and the rules governing a resurvey bring into consideration in a more comprehensive manner the position of recognized land boundaries, in the absence of evidence of the original corners. The surveyor has noted the detailed instructions set forth in Chapter V looking to the identification of existing evidence of the original survey and the application of the rules of proportionate measurement for the determination of the theoretical position of lost corners. These rules will be applied in the dependent resurvey generally with respect to the township as a unit, wherein the means of identification of each and every existent corner will be exhausted and the theoretical position determined for each lost corner. The former are to be considered as fixed points (except in most unusual cases) and may be monumented at any time; the latter will be subjected to the possible influence of points which may afterwards be determined to be acceptably located under the same rule of good faith, and will be marked only as temporary points until this question has been disposed of.

412. A complete retracement of the original survey will be made, based upon known corners, it being assumed that the exterior boundaries of the township to be resurveyed have been identified or restored under the rules already laid down in Chapter V, and under those relating to the acceptability of a local point or claim location. It is not usually possible to follow the method and order of procedure shown in the record of the original survey (owing to missing corners), but the complete system of lines will be run out by preliminary retracement, usually beginning with the meridional lines between known corners, followed by the latitudinal lines between known corners, noting the intersections with the said meridional lines. The surveyor must be supplied with a complete copy of the record of the original survey, and temporary reference stakes may be set on the meridional lines at the record measurement for each corner point.

413. The preliminary retracements will lead at once to the identification of the prominent evidence of the original survey and a trial calculation will follow as to the latitudinal and longitudinal adjustments at each missing corner, to suit the proportions which may be derived when based upon these known corners. A second and more exhaustive search will then follow within the zone of the probable location of each missing corner for the more obscure evidence of the original survey. At this stage of his field work the surveyor should exhaust every possible means of identifying the existent corners of the original survey. In many respects, the surveyor will be compelled to devise his own methods as the actual field conditions seem to warrant, and his skill and judgment as a surveyor should function to the fullest capacity.

If additional evidences of the original survey are found by this process, a second trial calculation will then be made as to the latitudinal and longitudinal adjustments of the temporary reference stakes previously set at each missing corner, to suit the proportional measurements derived from all of the known original corners—exactly as outlined in Chapter V. These calculated adjustments will determine the theoretical location of each lost corner with reference to all existing evidence of the original survey.

In the absence of other considerations, the theoretical points thus determined by proportionate measurement, based upon existing original corners, are fixed to a mathematical certainty, and when these points have been determined, the evidence of the original survey and the record thereof have served their primary purpose. Then, and not until that time, is the surveyor prepared to consider the weight of such collateral evidence as may be available.

**414.** The question now to be determined is whether the position of the lands claimed, occupied or improved is to be adopted under the rule of good faith as to location, and whether, if so adopted, the claims thus acceptably located can all be properly protected by the dependent plan of resurvey. If the position of any claim fails to qualify under the said rule of good faith it may be disregarded as to the effect produced thereon by the plan of dependent resurvey. On the other hand, if these claims are held to be acceptably located under the same rule, they may be adopted as the determining factor in the position of the missing corner or corners; and if the claims are in such concordant relation to each other and to the identified evidences of the original survey as to receive full protection by the dependent plan of resurvey, the surveyor may proceed with full assurance of the adequacy of the plan. Otherwise, the question of other processes analogous to those of an independent resurvey (as hereinafter explained) must be considered.

If two or more claims are acceptably located, but are discordantly related to each other to a considerable degree (by virtue of irregularities in the original survey), it will be clear that the general plan of dependent resurvey may not afford protection to such claims; whereupon the influence thereof must be rejected in favor of the theoretical point previously determined by proportional measurement. In this case, as before stated, some other process must be adopted to protect the acceptably located claims.

**415.** These acceptably located points for the missing corners will receive all the authority and significance of an identified original corner, and when the influence thereof on the dependent plan of resurvey has been combined with that of the existing original corners previously identified, the latitudinal and longitudinal adjustments of the temporary points on the meridional lines may be made accordingly.

**416.** In cases of distortion, if the distorted lines are to be adopted in the plan of dependent resurvey, it should be remembered that the lengths of lines, when subject to double proportion, are comparable only when reduced to cardinal equivalents or to equivalents along the direct lines between the nearest existing corners.

**417.** Many situations will arise where it will be manifest to the surveyor that it is better to accept a position based upon local improvements rather than to disturb satisfactory existing conditions. The surveyor will endeavor to avoid disturbing the position of locally recognized lines when such action may adversely affect improvements, and at the same time extreme caution will be exercised in the matter of adopting local points of control, which when accepted must be given, as above stated, a significance similar to that of an original corner and be allowed to function on an equality therewith. The acceptance of duly qualified and locally recognized points of control should aid materially in obtaining simplicity of resurvey and avoid the need for special metes-and-bounds surveys (as hereinafter described), which would differ only slightly in position from the regular lines of the resurvey. In this manner a flexibility will be introduced in the application of a dependent resurvey, at least to the point of protecting satisfactory local adjustments.

**418.** The surveyor should fully understand that the field of influence to be exercised by any acceptable location must be restricted to that already covered in a larger way by the existing evidences of the original survey, and that the adjustive process is of more or less local application. In this connection, it should be noted that the record of the original survey can not be abandoned in favor of an *indiscriminate* adoption of property corners, all or a portion of which fail to qualify as aforestated, nor is it to be assumed that because a large number or all of the claims within a township are consistently related among themselves to an arbitrary system of control which is itself altogether unrelated to the original survey, that such system is necessarily to be adopted as the basis of a dependent resurvey.

**419.** Thus where bona fide rights, as defined hereinbefore, are found to have been definitely established with reference to the location of lands the position of which can not otherwise be fully demonstrated by existing evidence of the original survey, the surveyor engaged in the resurvey will reject the theoretical point determined by the primary control in favor of a near-by *duly qualified* corresponding point, the position of which has been agreed upon by the adjoining property owners. Such a point may be recognized as the best available evidence of the true position for a corner; as previously stated its acceptance by the surveyor confers upon the point a significance similar to that of an original corner position, and thus avoids disturbing satisfactory local adjustments. Chief among this class of evidence forming the basis of the recognized position of land boundaries are recorded monuments established by local surveyors, duly agreed upon by the interested property owners; the position of boundary fences determined in the same manner; and the center lines of public roads and drainage or irrigation ditches, when intended to be located on the subdivisional lines of the public-land surveys. The local record in these cases, when available, may furnish the connecting link to the previously identified evidence of the original survey, but even in the absence of a conclusive record, if a point qualifies as above outlined, the presumption is strong that its position bears satisfactory relation to the original survey and that its correctness can not be successfully disputed. Points which actually qualify as aforestated may be accepted as the best available evidence of the true position of the original survey.

**420.** The technical record of the resurvey should clearly set forth the reasons for the acceptance of a local point, where unofficial determinations of the above character do not represent actual marks of the original survey. Such recognized and acceptable local marks will be preserved, and described in the record of the resurvey. New monuments will be established as required, in addition to, but without destroying the evidence of, the local marks.

### REESTABLISHMENT OF TRUE LINES.

**421.** As already stated, with the combined control of the dependent resurvey fully determined, the final calculation will be made as to the latitudinal and longitudinal adjustments of the temporary reference stakes previously set at the remaining missing corners. The final calculations will be based upon the known position of the corners of the general control as thus adopted, upon the plan of proportionate measurement, all as provided in Chapter V. The result of this process balances in regular proportion the differences between the measurements shown in the record of the original survey and those derived in the retracement. Thus the true lines of the dependent resurvey are finally determined through the influence exercised by the identified existent corners of the original survey and every other identified call of the record thereof, and such other collateral evidence of the position of recognized land boundaries as may be properly adopted for such influence.

**422.** The field procedure incident to the running and measurement of the true lines of the dependent resurvey will conform to the requirements of Chapter II, while the marking of lines between corners and the notation of objects to be recorded will conform to the provisions of Chapter III, and the monumentation of the survey will comply with Chapter IV. The technical record of the resurvey will be broadened to show the relationship between the original survey and its reestablished lines.

**423.** The field note description of an identified or accepted corner will be introduced into the technical record of the resurvey at the place in the true line notes where the position for the corner is indicated as having been attained. The record will embrace:

(*a*) A complete description of the remaining evidence of the original monument;

(*b*) A complete description of the new monument;

(*c*) A complete description of the original accessories as identified;

(*d*) A complete description of the new accessories;

(*e*) A concise statement relating to the recovery of a corner based upon identified line trees, blazed lines, items of topography, or other calls of the field notes of the original survey, in the absence of evidence of the monument or its accessories; and,

(*f*) A statement of fact relating to the relocation of an obliterated monument; or a statement of the determining features leading to the acceptance of a recognized local corner.

**424.** General titles (in addition to the regular page heading) will be inserted in the field notes of dependent resurveys to indicate the character of the resurvey, the technical record of which follows. Such titles will be inserted in the body of the field notes, as appropriate, and will show the name of the original surveyor and the year in which the original survey was executed; as, for example:

"Reestablishment of the surveys executed by John B. Smith, U. S. Surveyor, in 1842,"

and additional memoranda will be added as appropriate, explanatory of the method of control adopted in the restoration of one or more lost corners.

**425.** In addition to the usual showing of data upon the township plat, the plat of a dependent resurvey should carry a memorandum for the information of the public to the effect (modified as special circumstances may warrant) that—

"This plat of the resurvey of T. ——, R. ——, delineates a retracement and reestablishment of the lines of the original survey as shown upon the plat approved ——— (date), in their true original position according to the best available evidence of the position of the original corners; all differences between the measurements shown on the original plat and those derived in the retracement have been distributed proportionally between accepted corners in accordance with surveying rules; reference will be made to the original plat for the showing of the areas and more detailed descriptions of the various smaller subdivisions."

### ADDITIONAL METHODS FOR THE PROTECTION OF BONA FIDE RIGHTS.

**426.** Referring to those cases where locally recognized corners are discordantly related to the original survey, it will be apparent that such corners can be employed only for the determination of the boundaries of claims where bona fide rights have been duly established which would otherwise be impaired by the resurvey under the same rule of good faith in the matter of location. Cases of this kind are found to be decidedly exceptional in the townships where dependent resurveys have been made, and such situations will be given particular attention in the preliminary examination and special instructions. In those instances when encountered, provision will be made in the special instructions for a "metes-and-bounds" survey, as hereinafter outlined under the general subject of "independent" resurveys, unless an amendment of entry in conformity with the lines of the resurvey will answer the particular requirements of the situation. In either case the surveyor will note the Manual text relating to metes-and-bounds surveys and amendment of entries (see secs. 434 to 452, inclusive).

### EXAMPLE

**427.** A hypothetical example of a dependent resurvey follows in the text, wherein a showing of typical conditions will be presented. In this connection it will be observed that the application of the rules for the execution of a dependent resurvey is generally made with respect to the township as a unit. In this hypothetical case it is presumed that a sufficient number of original corners can be identified to enable the restoration of the township exteriors resulting in a satisfactory closure. Upon retracement of the interior lines, some evidence of the original survey is developed, also certain recognized and acceptable corners. All claims are found to be conformable.

The surveyor will proceed with the complete retracement of the interior section lines. In this process he will employ instrumental methods and make the measurements as provided in Chapter II. He will be guided by the suggestions given in Chapter V in regard to the search for evidence of the original survey, and beyond that he will devise his own methods in the search as the actual field conditions seem to warrant. Temporary reference stakes will be set where the original corners are not at once identified (though the use of local reference points will be unobjectionable). It will be assumed that a single system of reference stakes has been employed, as this scheme lends itself more readily to theoretical discussion, as well as practical utility in the field, and allows the utmost freedom as to the order in which the retracements are made.

Having completed the reestablishment of the township exteriors and the retracement of the interior lines, the surveyor will be concerned with the two primary considerations, heretofore discussed, which it is his duty to harmonize: First, the restoration of what the record purports to be original conditions; and, second, the protection of the bona fide rights of claimants in the matter of location. The first requirment must be fulfilled with reference to the evidence of the original survey, and the discovery and identification of actual original corners is paramount, bearing in mind that the development of a single additional original corner adds manifest conclusiveness to the work. These identified points when combined with those acceptably located constitute the general control. The second item, which does not directly affect the technical procedure, has been fully discussed hereinbefore.

### KEY TO DIAGRAM, FIG. 69.

A. Identified original corner.

B. Intersection of center lines of public crossroads, intended to be located at section corner and generally so recognized; accepted as best available evidence of corner.

C and D. Identified original corners.

E. Corner established by local surveyor; record shows proper application of the method of double proportionate measurement; generally recognized as correct position of corner; accepted on an equality with an identified original corner.

F–M, inclusive. Identified original corners.

N. Same as B.

O. Identified original corner.

P. Intersection of mean position of meridional and latitudinal blazed lines through virgin timber; age count on overgrowth qualifies for date of original survey.

Q. Restored corner based upon control furnished by latitudinal position of blazed line as above and fixed in departure by distance to original line tree.

R. Identified original corner.

**S.** Same as E.

**T.** Position determined by location of improvements; point agrees approximately with the theoretical position and it is recognized by adjoining claimants; improvements would be adversely affected by change of point.

**U.** Same as E.

**V** and **W.** Same as T.

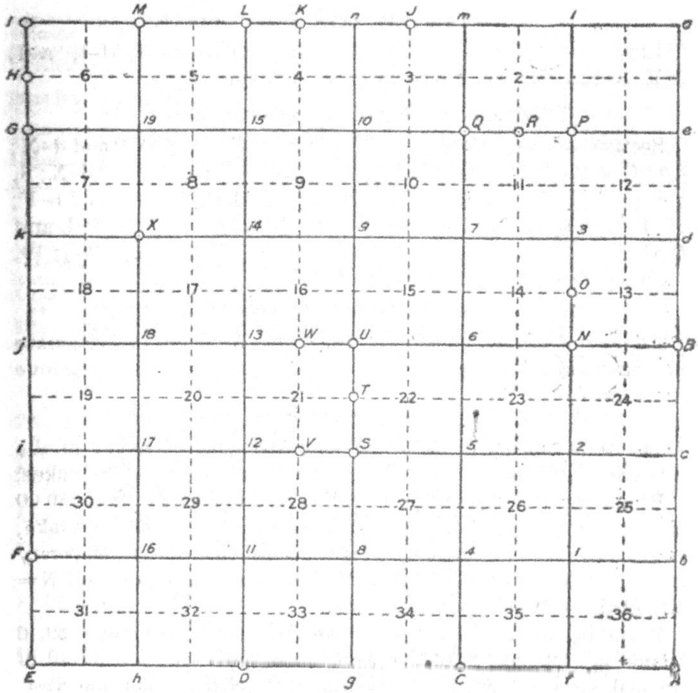


○ Employed for general control.
+ Theoretical position.

Fig. 69.

**X.** Identified original corner.

**a.** Duly restored by double porportionate measurement and thereafter employed for general control on an equality with an identified original corner.

**b–n, inclusive.** Theoretical true line position, duly restored by single proportionate measurement.

### METHOD.

After completing all retracements and having determined upon the general control to be adopted, as indicated in the diagram and accompanying key, the true lines of the dependent resurvey, beginning at the southeast corner of the township, will be reestablished as follows:

### SINGLE PROPORTIONATE MEASUREMENT.

Lines: A–B, B–a, A–C, C–D, D–E, E–F, F–G, G–H, H–I, a–J, J–K, K–L, L–M, and M–I.

### DOUBLE PROPORTIONATE MEASUREMENT.

Section corners: 1, f–N and b–F; 2, f–N and c–S; 3, O–P and d–X; 4, C–Q and b–F; 5, C–Q and c–S; 6, C–Q and N–U; 7, C–Q and d–X; 8, g–S and b–F; 9, U–n and d–X; 10, U–n and Q–G; 11, D–L and b–F; 12, D–L and V–i; 13, D–L and W–j; 14, D–L and d–X; 15, D–L and Q–G; 16, h–X and b–F; 17, h–X and V–i; 18, h–X and W–j; 19, X–M and Q–G.

### INTERIOR QUARTER-SECTION CORNERS.

All missing interior quarter-section corners by single proportionate measurement on line between the adjoining section corners as above determined.

### FIELD DATA.

The retracements develop the following data in regard to the relative position of certain points of control and the temporary stakes:

Beginning at f, North, 40.00 chains, set temporary stake; 80.00 chains, set temporary stake; 120.00 chains, set temporary stake; 160.00 chains, set temporary stake; 200.00 chains, set temporary stake; 241.20 chains, fall 90 links W. of N.; meridional excess f–N= 1.20 chains=40 links per 80.00 chains.

Beginning at b, West, 40.00 chains, set temporary stake; 80.46 chains, fall 20 links N. of temporary stake previously set; record of original survey shows length of line 80.22 chains; continue west, etc., to F; latitudinal deficiency b–F=84 links=14 links per 80.00 chains.

Beginning at 2 (temporary stake), East, 40.00 chains, set temporary stake; 80.82 chains, fall 44 links S. of c; record of original survey shows length of line 79.90 chains; run west from temporary stake at 2 on similar plan; latitudinal excess c–S=66 links=22 links per 80.00 chains.

### CALCULATIONS.

The adjustments of the temporary stakes to true line position, and the determination of the bearings and lengths of the reestablished true lines, are calculated as follows:

BETWEEN SECTIONS 35 AND 36.

| Memo. | Course. | Distance. | N. | S. | E. | W. |
|---|---|---|---|---|---|---|
| Retracement f-1 | North. | 80.00 | 80.00 | | | |
| Adjustment at 1 for meridional excess. | | | .40 | | | |
| Adjustment at 1 for latitudinal deficiency, 80.46—(80.22—0.14). | | | | | 0.38 | |
| True line f-1 | N. 0° 16′ E. | 80.40 | 80.40 | | .38 | |
| Adjustment at f | | | .00 | | .00 | |
| Adjustment at 1 | | | .40 | | .38 | |
| Adjustment at ½ sec. cor. (mean) | | | .20 | | .19 | |

BETWEEN SECTIONS 25 AND 26.

| Memo. | Course. | Distance. | N. | S. | E. | W. |
|---|---|---|---|---|---|---|
| Adjustment at 1 from true to temporary. | | | | 0.40 | | 0.38 |
| Retracement 1-2 | North. | 80.00 | 80.00 | | | |
| Adjustment at 2 for meridional excess. | | | .80 | | | |
| Adjustment at 2 for latitudinal excess, 80.82—(79.90 +0.22). | | | | | 0.70 | |
| | | | 80.80 | .40 | .70 | .38 |
| | | | .40 | | .38 | |
| True line 1-2 | N. 0° 14′ E. | 80.40 | 80.40 | | .32 | |
| Adjustment at 1 | | | .40 | | .38 | |
| Adjustment at 2 | | | .80 | | .70 | |
| | | | 1.20 | | 1.08 | |
| Adjustment at ½ sec. cor. (mean) | | | .60 | | .54 | |

BETWEEN SECTIONS 23 AND 24.

| Memo. | Course. | Distance. | N. | S. | E. | W. |
|---|---|---|---|---|---|---|
| Adjustment at 2 from true to temporary. | | | | 0.80 | | 0.70 |
| Retracement 2-N | North. | 81.20 | 81.20 | | | |
| Random line to N | East | .90 | | | 0.90 | |
| | | | 81.20 | .80 | 90 | 70 |
| | | | .80 | | .70 | |
| True line 2-N | N. 0° 9′ E. | 80.40 | 80.40 | | .20 | |
| Adjustment at 2 | | | .80 | | .70 | |
| Adjustment from 80.00 ch. point on random to N. | | | 1.20 | | .90 | |
| | | | 2.00 | | 1.60 | |
| Adjustment at ½ sec. cor. (mean) | | | 1.00 | | .80 | |

BETWEEN SECTIONS 25 AND 36.

| Memo. | Course. | Distance. | N. | S. | E. | W. |
|---|---|---|---|---|---|---|
| Retracement b-1 | West. | 80.46 | | | | 80.46 |
| Random line to temporary stake at 1. | | | | 0.20 | | |
| Adjustment at 1 temporary to true. | | | 0.40 | | 0.38 | |
| | | | .40 | .20 | .38 | 80.46 |
| | | | .20 | | | .38 |
| True line b-1 | N. 89° 51′ W. | 80.08 | .20 | | | 80.08 |
| Adjustment at b. | | | | .00 | | .00 |
| Adjustment from 80.00 ch. point on random to temporary stake at 1. | | | | .20 | | .46 |
| Adjustment at 1 temporary to true. | | | .40 | | .38 | |
| | | | .40 | .20 | .38 | .46 |
| | | | .20 | | | .38 |
| | | | .20 | | | .08 |
| Adjustment at ¼ sec. cor. (mean). | | | .10 | | | .04 |

BETWEEN SECTIONS 24 AND 25.

| Memo. | Course. | Distance. | N. | S. | E. | W. |
|---|---|---|---|---|---|---|
| c to random line | | | | 0.44 | | |
| Retracement c-2 (reversed) | West. | 80.82 | | | | 80.82 |
| Adjustment at 2 temporary to true. | | | 0.80 | | 0.70 | |
| | | | .80 | .44 | .70 | 80.82 |
| | | | .44 | | | .70 |
| True line c-2 | N. 89° 45′ W. | 80.12 | .36 | | | 80.12 |
| Adjustment from 80.00 ch. point on random to c. | | | .44 | | .82 | |
| Adjustment at 2 temporary to true. | | | .80 | | .70 | |
| | | | 1.24 | | 1.52 | |
| Adjustment at ¼ sec. cor. (mean). | | | .62 | | .76 | |

## THE INDEPENDENT RESURVEY.

**428. An independent resurvey is an official re-subdivision of the public lands distinct from the original survey which it is designed to supersede. The independent resurvey is accomplished by three distinct steps:**

**(*a*) The reestablishment of the outboundaries of the lands subject to resurvey, following the method of a dependent resurvey;**

(*b*) The segregation of lands embraced in any valid claim where the initial steps have been taken looking to the disposal of the title of the United States based upon the former approved plat; and,

(*c*) New exterior, subdivisional and meander lines as necessary, established upon a new regular plan, which, for every purpose of identification and description of the public lands involved, becomes the prevailing survey.

REESTABLISHMENT OF OUTBOUNDARIES.

**429.** The limiting boundaries of the lands subject to independent resurvey must agree with the previously established and identified exterior or subdivisional lines of the approved original surveys. In order to qualify as a suitable limiting boundary as aforementioned, a line of the accepted established surveys will be selected which can be conclusively identified (by existing original or properly restored corners) in one position to the exclusion of all others and which by its known position adequately protects all rights (located in good faith as hereinbefore defined) based upon any township plat showing subdivisions of the public lands adjacent to said boundary. Such outboundaries of the lands to be resurveyed by the independent process must necessarily be retraced and reestablished in their true original position. The lands upon one side of such outboundary are to be re-subdivided upon a new plan, while upon the opposite side of such line the original subdivisions are to be strictly maintained and none of the original conditions are to be disturbed.

**430.** The outboundaries are generally selected along the locus of the previously established township exteriors where the existing evidence gives positive proof of the location of the original survey, and where conditions on the ground are harmoniously related to the record of said original survey. In special cases certain section lines may fully qualify as suitable lines to mark the limit of the independent resurvey; such section lines will then be duly retraced and reestablished in their true original position. Particular attention will be given to this very important subject at the time when the field examination is made with a view to maintaining the original survey as far as consistent.

**431.** In those cases where a proper limiting boundary can not be secured without involving the necessity for the inclusion in the group of a greater number of townships than administratively practicable to execute in one assignment, the necessity may arise for the extension of tract segregations (as hereinafter outlined) into a township ungrouped for resurvey. In such cases, under specific authority of the General Land Office, any tract found to extend across such group outboundary will be segregated in full, whether or not the tract was originally described as in the township to be resurveyed, and the necessary steps will thereupon be taken by the General Land Office in the matter of suspension of the lands in the adjoining township from further disposal and of additional investigations with a view to a resurvey of all or a portion of the said adjoining township. (See second rule, sec. 445.)

**432.** The special instructions will show specifically what lines have been selected to limit the independent resurvey, and the surveyor engaged in the execution of such resurvey will proceed with the retracement and reestablishment of said outboundaries as a condition precedent to beginning the independent resurvey.

**433.** Where the new lines of the independent resurvey are not to be initiated or closed upon the restored original corners of the reestablished outboundaries of the independent resurvey, said restored corners will be marked only with reference to the township, range and section to which they will thenceforth relate, and new regular corners of minimum control will be established as necessary to govern the lines of the independent resurvey, all as provided in sec. 164, Chapter III. During the preliminary stages of the resurvey there will often be more or less doubt as to whether an old corner will retain its former control or will have to be altered, and until this uncertainty has been removed the marking of a corner and its accessories should be deferred. The monumentation will follow the final determination of the future significance of each point. Where an old point is to be perpetuated merely to control the former alinement, but not the corner of a subdivision, its future significance will be that of an "angle point" only and the monument and its accessories will be marked accordingly.

METES-AND-BOUNDS SURVEY OF PRIVATE CLAIMS.

**434.** After the reestablishment of the outboundaries of the lands subject to independent resurvey has been accomplished in accordance with the requirements of the special instructions, the surveyor's attention will be directed to the segregation or marking out of all duly entered, selected, reserved (in certain cases), granted, or patented lands whose description may be based upon the former approved plat, and which can not be conformed to the lines of the resurvey.

435. A status diagram will be furnished to the surveyor showing all patented lands, valid entries, school sections, and other land grants, and all other disposals, reservations, or selections of lands whose position and description are based upon the original survey and plat, and whose boundaries can not legally be disturbed. In every case the various tracts shown upon the status diagram will be protected either by individual "metes-and-bounds" survey or by the assignment of appropriate subdivisions of the resurvey in case the latter lines (new section lines, or center lines of sections or quarter sections) are found to coincide or approximately agree with the boundaries of said tracts.

436. It is not to be understood that the metes-and-bounds survey of private claims must be completed before beginning the projection of the new lines of the independent resurvey. It has merely been deemed logical to consider the subject of the tract segregations in advance of the question of the establishment of new lines. The fact is that surveyors will find it expedient to carry both branches of the survey along together in the locality of the camp or other field headquarters.

437. The jurisdiction of the General Land Office, the limit of the authority of the surveyor, and the bona fide rights of claimants, where entered or patented lands are involved, remain absolutely the same whether the resurvey is to be made upon the dependent or independent plan. Thus where the independent type of resurvey has been adopted as more feasible, identified corners of the original survey in the immediate vicinity of lands to be segregated will be employed for the control of the location of such lands. The question of the good faith of the entryman will in every case be fully considered, as previously outlined in this chapter, and where the evidence of the original survey is so obliterated that a charge of a lack of good faith can not be brought against an entryman whose claim boundaries may differ from a theoretical location determined by more rigid surveying rules, the position of the improvements is to be regarded as the best available evidence of the original position of the claim, and the same will be employed as far as consistent for the control of the location of the boundaries of such claim.

438. Where there is sufficient evidence of the original survey, the identification of the areas to be segregated, resulting from the sub-

division of the original sections, will proceed in accordance with the provisions of Chapters III and V, and every corner or angle point of each tract as thus located will be marked upon the ground.

**439.** Where the surveyor can not point out, by suitable identification of the original surveys, the definite location of an entry based upon the former approved plat, the claimant or owner of such lands will be consulted as to the position of his boundary lines. The boundaries of the private claim, so determined, will be fixed, as between the private and public lands, subject to the official acceptance of the resurvey. Where dispute is encountered in regard to the adjustment of the line between adjoining patented tracts, each acceptably located under the rules already laid down, which can not be reconciled or suitably disposed of by surveying process, the tracts will be surveyed in conflict, as hereinafter provided, and so shown on the resurvey plat; the questions arising out of such conflict will be given administrative review with the field notes of the resurvey.

**440.** The owner of an unidentified claim will be called upon to indicate the boundary lines thereof if possible, and in this connection, should occasion arise, the surveyor will explain the manner of adjusting differences between adjoining claims and what will constitute an acceptable location of a claim. The latter condition demands a form agreeing with the original entry, approximately regular boundaries, an area not widely inconsistent with that shown upon the original plat, and a location as nearly correct as may be expected from the existing evidence of the original survey, without overlapping into an adjoining township not subject to resurvey, except as provided in sec. 431. In every case where the outboundaries of the lands subject to "independent resurvey" have been reestablished by the "dependent" or "restorative" plan, the subdivisions of a tract situated and originally described as along or upon the opposite sides of such outboundary must agree with the line reestablished and harmonize in relative position.

**441.** In the execution of an independent resurvey, therefore, the identity of each tract to be segregated therein or indicated by conformation to the lines of the resurvey, whether patented or unpatented, must be maintained, and the surveyor will not be allowed to change materially the configuration of a tract as shown by its original description in order to indemnify the owner thereof against deficiencies in area, to eliminate conflicts between entries, or for any other purpose. If improvements have been located in good faith, the segregation survey should be so executed, or the conformation to the lines of the resurvey so indicated, as to cover as nearly as possible these improvements and at the same time maintain substantially the form of the entry as originally described. No departure from this rule will be allowed.

**442.** The question of amendment of entries for the purpose of permitting adjustments in terms of the resurvey involving lands not included within the original tract is a matter for the adjudication of the General Land Office after the resurvey has been accepted and the plats thereof filed in the local land office.

**443.** In case of absentee owners an attempt should be made to establish communication, if necessary, in order that the claimant may point out the lands subject to a metes-and-bounds survey. If the owner can not be found and there is no visible indication, such as a boundary fence, of the location of the limits of a claim, the surveyor will exercise the alternative of locating the claim from the nearest original point of control or from a point of a neighboring claim, or of assigning to the entered or patented lands the appropriate subdivisions of the resurvey, all subject to the principles hereinbefore set forth. The controlling factors in such locations will be based upon the individual and neighborhood improvements (such as buildings, wells, springs of water, cultivated lands, public roads, fences, corners of recognized private surveys, etc.) which may indicate the evident intention of the entryman or patentee as to the position of his land.

**444.** Each non-conformable valid claim in a township will be given a serial tract number, commencing with No. 37 in the smallest numbered and entered section of the original plat, progressing through the township in the order in which lots and sections are numbered. A tract number will be used but once in a township, and if any tract lies partly in two or more townships subject to resurvey the number applied to the tract in the first township resurveyed will not be used for other tracts in the adjoining township.

**445.** The following rules will be observed in the execution of the metes-and-bounds survey of all specially designated tracts:

1st. Each claim, acceptably located, but at variance with the lines of the resurvey, will be surveyed and monumented at each angle point.

2d. Where a portion of a claim is originally described as in a township not subject to resurvey, such portion of the claim will not be surveyed by metes and bounds, provided the limiting boundary is found to qualify as set forth in sec. 429. The portion of the claim originally described as in the township to be resurveyed should ordinarily be defined in a position (either by segregation or conformation to the lines of the resurvey) which is properly related to the identified or restored corners on the limiting boundary. (See sec. 431.)

3d. Where the boundaries of a claim are unacceptably located as pointed out by the claimant, the surveyor will proceed with a proper survey of the tract in accordance with rules already stated which will result in a suitable relation to the original survey, and the corners of the tract as thus located will be monumented. If the claimant protests against such location, the surveyor will request that the protest be made in writing (to be submitted with the returns of the resurvey), and will thereupon make an accurate connection with the corners of the claim as unacceptably located, to be made the subject of a complete report by the surveyor in his field notes, reviewing the facts with reference to the question of location. As a further protection to an entryman thus unacceptably located see sec. 455.

4th. Where, through a compliance by the surveyor with the general rules above laid down, the metes-and-bounds segregation of a claim (or the conformation thereof to the lines of the resurvey) within the field of an independent resurvey (or the related subdivisions within the field of a dependent resurvey) fails to cover any or all of the lands, occupied, improved or claimed by the entryman, patentee or present owner, and the latter indicates a desire to amend his entry, a full report will be made by the surveyor in his field notes, describing therein the subdivisions actually occupied and sought to be acquired under the amended entry, but which are not covered by the tract as surveyed, all looking to the protection of the title to the lands actually earned.[1]

5th. Where it so happens that the regular quarter-quarter sections embraced within a claim fall in approximately the same position as the regular quarter-quarter sections of the resurvey, and the entryman or patentee indicates a desire to conform his claim to the resurvey, and no apparent objection is found by the surveyor, the facts will be stated in the field notes, and the claim will be so indi-

[1] See current circular relating to amendment of entries.

cated upon the resurvey plat. Under this circumstance the metes-and-bounds survey of the tract will be omitted. However, where any tract whose original description includes *any fractional lot*, or where any part of a tract falls upon *any fractional lot* of the resurvey, the tracts will be segregated as a whole by metes-and-bounds survey, even though some or all of the lines of the tract may coincide with certain subdivisional lines of the resurvey.

6th. Conflicting tracts, each acceptably located, will be surveyed and monumented, and conflict shown upon the resurvey plat. Each intersection of conflicting boundaries will be determined upon the ground and recorded in the field notes.

7th. The angle points of a tract will be designated by serial numbers beginning with No. 1 at the northeast corner, and proceeding around the claim, running westerly from the initial corner. An angle point may be common to one, two, three or four tracts, and will be monumented and marked as provided in Chapter IV; as for example:

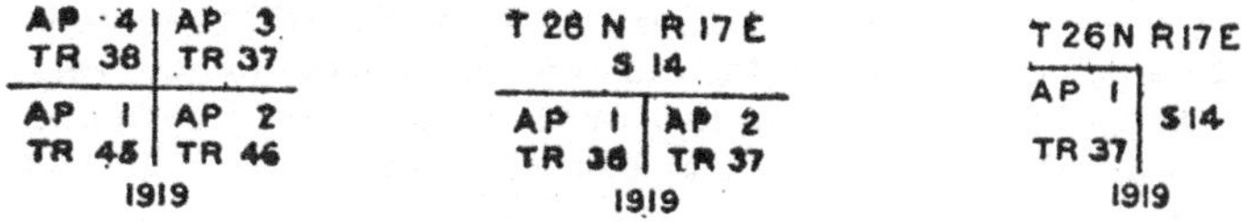


8th. No accessories will be required with the monuments at the angle points of the metes-and-bounds survey.

**446.** The proper supervising officer will furnish the surveyor with an abstract of the valid entries, selections, reservations, patents, and grants, based upon the original plat of any township (or portion thereof) subject to resurvey, and the said resurvey can not be regarded as complete until each and every claim described in said abstract of entries (and shown on the status diagram) as in the township to be resurveyed has received full protection in the matter of location. Aside from those disposals described as in the township to be resurveyed, there will also be furnished to the surveyor, as a matter of information, the status of all claims in the adjacent sections of all adjoining townships ungrouped for resurvey. The abstract will be included with the other data to accompany the written special instructions providing for a resurvey.

**447.** The field notes of the metes-and-bounds survey of each valid claim will be preceded by a copy of the abstract of entry thereof. A brief statement will then follow in each instance (or with suitable reference), concerning the principal factors controlling the location of the particular tract, and whether or not the claimant was consulted, or communicated with, in the matter of the identification of the boundaries of his claim. The statement should be clear as to whether the location of a claim, shown either as a tract segregation or as conforming to the lines of the resurvey, was controlled by improvements alone, or by one or more identified corners of the original survey, nearby or remotely located, or by its relation to adjoining tracts. In case all of the tract segregations within a township can be covered by one general statement, the same should appear at the beginning of the field notes of the metes-and-bounds surveys. The field notes should be made to account for each and every tract shown upon the status diagram.

**448.** All claims should be accounted for on the resurvey plat, and all will be shown either as segregated tracts or as conforming to the lines of the resurvey, as the case may be, with outline indicated by heavy black lines. An exception to this rule will be made in those rare cases where all the claims within a township have been conformed to the lines of the resurvey *under their original description*, in which event a statement may be made on the margin of the plat that—

"All claims originally described as in this township are intended to conform to the lines of the resurvey under their original description."

**449.** As a further safeguard that the returns of independent resurveys may be conclusive in the matter of the significance of the tract segregations, the plats thereof will show a statement that—

"All tract segregations shown hereon represent the position and form of said tracts under the original description as referred to the original survey, located as such on the ground according to the best available evidence of their true position."

**450.** The above statement will be modified if one or more of all the claims shown on the status diagram are conformed to the lines of the resurvey, either under the original description or by different legal subdivisions, as follows:

"All tract segregations shown hereon and all other claims shown to conform to the lines of the resurvey, whether by the original or new legal subdivisions, represent the position and form of said tracts under the original description as referred to the original survey, located as such on the ground according to the best available evidence of their true position."

451. The projection and measurement of the lines of the metes-and-bounds survey and the technical record in respect to the same will conform to the usual practice in regular surveys. While the mapping of important items of topography and valuable permanent improvements will be given attention with regard to this feature of the resurvey plat, yet it will be apparent that the amount of data to be shown in connection with the metes-and-bounds surveys makes it impossible, at the usual scale, to show objects of little relative importance. This class of memoranda taken during the progress of the work will *not* be required in the field notes of metes-and-bounds surveys.

452. At least one angle point of each tract survey will be definitely connected with one of the regular corners of the resurvey, and where lines of claims are intersected by lines of the resurvey a connection will be made from the point of intersection to the nearest claim corner and recorded in the field notes of the regular section line. The latter will be considered a satisfactory connection to all adjoining claims located within the interior of either section. Where an extensive system of tract segregations has been surveyed, the interior tracts of the block will not require individual reference connections. The establishment of closing corners on the regular line when entering or leaving public land will conform to the general practice in this respect as provided in sec. 191, Chapter III.

THE PROJECTION OF NEW LINES.

453. The peculiar conditions of the situation which necessitate an independent resurvey render it impossible to formulate general rules suited to all cases. Experience has demonstrated the necessity for giving deliberate attention to the unique problems of subdivision which are to be found in each definite example. The general practice is to secure a surveyor's report of the actual conditions involved in a particular independent resurvey, upon consideration of which there may be devised the best plan for a re-subdivision of the vacant public lands, and the latter will be set forth in the special instructions. The possibility of placing the regular lines of the independent resurvey so as to obtain maximum agreement with the position of the boundaries of conformable claims will be fully considered with a view to eliminating or reducing the necessity for tract segregations, if possible, where this can be accomplished in harmony with the rules previously outlined. The examiner's

recommendations in these matters should be explicit and responsive to his special advantages in the opportunity of working out the technical problem while on the ground.

454. A problem involving the re-subdivision of vacant public lands, as in an independent resurvey, should be approached in the same way as practically all problems in fragmentary subdivision, though the independent resurvey may at times involve the re-subdivision of a group of many townships wherein all conditions, except perhaps with relation to the tract segregation surveys, may be comparatively regular. First attention will be given to completing the new township exteriors which are to be independently resurveyed after having reestablished the outboundaries of the group on the dependent plan. The new exteriors will be carried forward and completed in harmony with the rules set forth in Chapter III for the establishment of original surveys. The new section lines will be run out and marked as in regular or fragmentary subdivision as the situation may be and new meander lines will be run as required. The new exterior and subdivisional lines will usually be extended across small blocks of tract segregation surveys, noting connections as previously stated, and in such cases the new lines and corners will be fully monumented regardless of the fact that some points will fall within the tract segregation surveys. The latter points are required in their usual function to determine the subdivision of the public lands affected.

455. A general exception to the rule of extending the lines of the independent resurvey across the tract segregations will be made in those townships or portions thereof so densely covered by private claims that the remaining parcels of public lands may be as well or better identified and described for expediency with reference to isolated tract numbers. In such cases closing corners will be required on the regular lines when entering or leaving public land. The regular lines may or may not be extended as blank lines across the tract segregations, according to the plan of running the new section lines of the resurvey. Where this method is employed it will be necessary to assign tract numbers to the vacant parcels of public land and to mark the angle points thereof accordingly. Where a parcel of vacant public land is to be identified on this plan, such vacant tracts will be surveyed by metes and bounds in accordance with the usual rules. Rare cases may arise where it will be deemed expedient to segregate by metes-and-bounds survey certain quarter-

quarter sections of vacant lands in accordance with the system of the original survey as indicated by adjoining tract segregations for the purpose of affording a better basis of disposal or for amendment of entries. Such segregations will not be made unless it is conclusively shown by the surveyor that the fractional lots and regular quarter-quarter sections of the resurvey are inadequate as a basis of disposal under existing conditions of occupancy on the part of settlers or of entrymen who may propose to amend. The special instructions will be made as explicit as possible in these details, which will be determined upon when the plan of the resurvey is under consideration by the supervising officer.

**456.** Where a section of the resurvey is invaded by patented tract segregations, but not by unpatented entries or selections, the lotting of the public lands will be carried out in accordance with the usual plan of lotting within fractional sections as outlined in Chapter III. The numbering of the fractional lots will begin with the number next higher than the highest number employed in the section of the original survey which bears the same township, range and section number. This plan is intended to avoid any possible confusion which might arise from a duplication in the use of the same lot numbers.

**457.** A departure from the usual rule for lotting is necessary in order to provide suitable descriptions within unpatented entries and selections where such tract segregations may be subject to relinquishment or cancellation, also in other cases, to facilitate a subdivision of isolated tracts of public lands surveyed by metes and bounds. Two methods have been found available, each one better suited to particular situations. Neither method involves any change in the instructions for the field procedure heretofore laid down. The discussion of the merits of the two methods and the examples of their use are better adapted to the text of Chapter IX, where the subject will be found in connection with other details to be shown upon the resurvey plats.

**458.** The general requirements of Chapters II, III and IV will be fully observed in every respect throughout the execution of the independent resurvey and in the technical record thereof. General titles (in addition to the regular page heading) will be inserted in the field notes to indicate clearly the character of the independent resurvey, the technical record of which follows; such titles will be inserted in the body of the field notes, as appropriate, and will show the full significance of all lines; as for example:

(a) "Metes-and-bounds survey of private claims as originally located in accordance with the survey executed by John B. Smith, U. S. Surveyor, in 1842;" and

(b) "Independent resurvey, superseding the survey executed by John B. Smith, U. S. Surveyor, in 1842."

**459.** All monuments of the original survey, not otherwise reported upon, when traces thereof have been found, will be connected by course and distance with a corner of the resurvey, and such connection and a description of the traces of the original corner as identified will be recorded in the field notes of the resurvey. A useless monument will be destroyed after the point is found to be no longer needed for the survey of a claim of any kind whose location may in any way depend upon such monument. (See sec. 163, Chapter III.)

**460.** Further exemplification of the approved practices incident to the successive field steps and preparation of the field notes and resurvey plats will be found in the chapters that follow.

O

Made in the USA
Monee, IL
04 November 2026

67385392R00177